Contents

Acknowledgements

This publication of the IPPR Media Project was made possible by the support of our sponsors, and in particular a generous award from the Gulbenkian Foundation.

The following organisations supported the IPPR Media Project in 1999-2000: BECTU, BBC, BSkyB, Independent Television Commission, Commercial Radio Companies Association, PACT, NTL, Channel 4, The Radio Authority, Turner Broadcasting, Panasonic , Kingston Communications.

Numerous other people have made this research possible, including volunteers, researchers and the permanent staff of the IPPR. I would like to thank Leslie Ross, Frank Jordans, Clare Heywood, Frank Poulsen and Jacob Dencik. I am particularly grateful to Seeta Gangadharan, for editing work on the texts.

The views expressed in the publication remain the responsibility of the authors, not necessarily those of ippr, its trustees or the sponsors of the project.

About the authors

Damian Tambini joined IPPR in February 1999 to take up the appointment of Senior Research Fellow. He was previously Fellow and Lecturer at Humboldt University in Berlin. He has been a visiting fellow at Nuffield College Oxford, The University of Westminster, The European University Institute in Florence and a researcher at the Institute of Education, University of London. His research interests include media and telecommunications policy and democratic communication. His publications include *Cyberdemocracy* (Routledge 1998) and *Markets, Citizens and States* (Oxford University Press 2001).

Liz Forgan OBE was formerly Director of Programmes at Channel Four and Managing Director of the BBC Network Radio. She is currently Director of the Most Media consultancy, as well as Chairman of Arrowhead Productions Ltd and a non-executive director of the Guardian Media Group.

Clare Hall was previously Research Fellow in UK Regulatory Policy and Administration at LSE, Research Officer at Bath University and the

COMMUNICATIONS
Revolution and reform

by Damian Tambini
with Liz Forgan, Clare Hall
and Stefaan Verhulst

ippr

30-32 Southampton St
London WC2E 7RA
Tel: 020 7470 6100
Fax: 020 7470 6111
postmaster@ippr.org.uk
www.ippr.org
Registered charity 800065

The Institute for Public Policy Research is an independent charity whose purpose is to contribute to public understanding of social, economic and political questions through research, discussion and publication. It was established in 1988 by leading figures in the academic, business and trade-union communities to provide an alternative to the free market think tanks.

IPPR's research agenda reflects the challenges facing Britain and Europe. Current programmes cover the areas of economic and health policy, governmental reform, human rights, social policy, the environment and media issues.

Besides its programme of research and publication, IPPR also provides a forum for political and trade union leaders, academic experts and those from business, finance, government and the media, to meet and discuss issues of common concern.

Production & design by **EMPHASIS**
ISBN 1 86030 138 X

University of Cambridge. Areas of expertise include: Regulation and organisations, Anthropology of policy, Knowledge management. Publications include *Telecommunications Regulation; Culture, Chaos and Interdependence Inside the Regulatory Process* (2000) London, Routledge.

Stefaan Verhulst has been the director of the Oxford University Programme in Comparative Media Law and Policy since 1996. Prior to that, he was a lecturer on communications law and policy issues in Belgium before becoming founder and co-director of the International Media and Info-comms Policy and Law studies at the School of Law, University of Glasgow. He has served and serves as consultant and researcher for numerous organisations including the Council of Europe, European Commission and Unesco. He became recently Chairholder - together with Monroe Price – of the Unesco Chair in Communications. He is the editor of the *International Journal of Communications Law and Policy* and the *Communications Law in Transition* newsletter and is the UK legal correspondent for the European Audiovisual Observatory.

Preface

Chris Smith MP

The transition to digital broadcasting, and the convergence of previously distinct technologies such as telephony, television and computing have been called a communications revolution. Whether current developments deserve that label only future historians will be able to tell us, but it is clear that the changes afoot represent a huge challenge to governments, and to communications companies, worldwide. It is crucial that the UK, a country that has long led the world in telecommunications and broadcasting regulation, takes the steps necessary to respond to the challenge.

Whilst we cannot yet ascertain exactly where current technological changes are taking us, we can now make out the general direction of change, and know that decisive action must be taken to reform the structure of communications regulation, remove regulation that is no longer necessary, and update that which remains necessary.

This publication, and the others that IPPR Media has produced in the past year have been an important part of the debate about how UK media regulation should respond to current changes during the process of the development of the Communications White Paper (available at www.communicationswhitepaper.gov.uk). The papers in this collection show that there is no final word in the debate about the digital future, but outline an approach that every participant in the discussion should note. The transition to digital is not just about protecting our media ecology, the creative, world-beating culture industries of the UK. It is also about developing a regulatory framework in which the UK can realise the positive potential of new communications technologies: to improve our democracies, create a more inclusive and democratic society, and a more effective and efficient information infrastructure.

Rt Hon Chris Smith MP
Secretary of State for Culture, Media and Sport

1. Introduction

Damian Tambini

The basic institutions and structures of broadcasting regulation in the UK emerged in the first half of the 20th century. The first private radio transmissions in the 1920s were quickly followed by the formation of the BBC. Other European countries also developed public broadcasting during the same period and monopoly public service provision regulated by strict codes for programming accountable to Parliament and the public became the dominant model. As new technologies emerged, and the number of domestically-owned receivers increased, the system evolved, but retained the standards and guidelines of public broadcasting. A limited number of licences to broadcast commercial TV and radio was issued. Because government controlled market entry through issuing licences to broadcasters, detailed content codes could be applied to govern what could and should be transmitted. Those who failed to comply would lose their licence to deliver broadcast standard pictures to the home.

The basic paradigm in regulation remained the same until the end of the century. The programme codes and standards for commercial broadcasting were similar to those in public monopolies. Despite periodic worries about capture of regulation by government or business, there was a broad consensus that regulation of content was necessary and justified, and that the UK enjoyed an exemplary system. Regulation in the UK reflected the idea that the peculiar nature and impact of broadcast communication justified detailed intervention in the public interest.

Many argue that the transition to digital will bring a fundamental departure from those basic structures of regulation. The approach to broadcasting regulation was based on an assessment of the characteristics of broadcasting as opposed to other forms of communication. Broadcast communications were seen as particularly public, influential, invasive and pervasive. Therefore, detailed regulation was justified. A clear distinction persisted between the ways that broadcast content and other content (phone-calls, print or tapes) were regulated. Broadcast content was seen as a matter for public and governmental concern, and was regulated 'in the public interest'.

The 'public interest' consisted in the following:

- Democratic communication: measures to limit media power and encourage pluralism;
- Positive content regulation: subsidies for educational, civic or cultural content;
- Negative regulation: restrictions on what was considered dangerous or harmful;
- Consumer protection and fair trading: regulating advertising and competition.
- Infrastructure: measures to ensure maximum access and utility in the networks.

In the final years of the 20th century, a new set of technological changes began to have an impact. At the start of 1998, very few households had digital television. By the end of 2000, more than 20 per cent of UK households had digital access, and projections supported the government's view that digital TV would be near universal by 2010. Further, it seemed that within the next decade, universal or near universal access to the Internet would be a reality. Indeed, achieving universal Internet access became a key government objective as it promised to bring widespread educational, social and democratic benefit. Whilst alternative forms of transmission, including satellite and cable, had long been increasing the number of channels – at least to those who could pay, the shift to digital and the rise of Internet protocol communication suggested a radical departure from the old model. The experience of the Internet seemed to prefigure a new world of any-to-any communications that frustrated the basic principles and structures of regulation in analogue broadcasting.

Old models of regulation were not well-adapted to the changing new media world for two basic reasons. First, in the new environment of multiple, niched services with smaller audiences, it no longer seemed justified to have the same degree of content regulation. Second, it no longer seemed possible given the volume and delivery means to combine with central regulation by detailed codes and oversight of all content. Internet enthusiasts predicted a future of wide range of services that high-speed Internet networks could make possible. For those who accepted this view, the role of government regulation of communications content was entering a rapid decline – the best thing it could do is stand aside.

Sceptics, however, held that this vision of digitopia was premature and unproven, that mass broadcasting remained the paramount source of information for most citizens and that detailed regulation therefore continued to be justified. They pointed out that the digitopia scenario was particularly convenient for the media players that had been long calling for deregulation of media markets. Viewing the media abundance argument differently, the sceptics argued that consumers were likely to stay with the mass, linear broadcast services, partly from habit and partly because of the undeniable economics of television. It costs a great deal to produce compelling, quality content. Even where new transmission facilities are available, people are primarily interested in using mass linear services. And given the intensifying trends towards vertical integration and concentration of media ownership, the sceptics argued, the traditional goals of regulation in the public interest for democracy and pluralism remain, and even intensify. For the digital sceptics there was no paradigm shift.

Whether a digisceptic or a digitopian, it is clear that new forces have to be taken into account during the transition to digital broadcasting. Technologies that deliver audio and video to the home are now becoming part of a broader network of content, data and information services. In the past, broadcasting was seen principally as a part of the cultural and educational fabric of the UK. Now in policy terms it is increasingly seen as a key asset of industrial infrastructure. Even opponents of new economy theory do not deny that connectivity is a major factor determining the industrial competitiveness of Europe and the UK, and the life chances of individual citizens. The goal of wiring up the country, and that of developing new forms of any-to-any communication that deliver public services, education, information and entertainment are now inextricably linked. Content regulation increasingly conflicts with industrial policy: whether it is the goal of public service broadcasting or negative regulation, the over-arching goal of rolling out the infrastructure and getting ahead in the global export of content is increasingly shaping the debate.

In the light of these developments, the UK government began planning a major piece of communications legislation. The government is to be commended for taking these issues head on, and promoting a broad debate, rather than simply accepting vested interest in the default position of the status quo.

> A principal goal for legislation is to facilitate rollout of the network infrastructure for cheaply distributing the large quantities of information that drive innovation and the new economy, and extending the benefits of that new dynamism to the largest possible number of people. But this process of reform has to be managed delicately. The same networks also carry the delicate ecology of UK broadcasting. Of at least equal importance in legislation and reform was to maintain and enhance the democratic, educational and cultural roles traditionally served by broadcasting in the new environment.

To its credit, the government made clear that the process of communications reform should not be a top-down imposition, but involve the widest possible consultation and debate. The papers gathered in this collection represent a small part of that debate, examining the key challenges for broadcasting regulation in the transition to digital. Would it be possible to square the circle and maintain the UK's extraordinarily high broadcast standards whilst spurring a race to digital? The chapters were discussed in a seminar series organised by the Institute for Public Policy Research in summer 2000, which gathered together media experts, academics, government advisors and representatives of the major UK broadcasters to outline the scope and principles of communications policy in the transition to digital. This collection does not offer a blueprint for the future. It does however attempt to take seriously some of the challenges currently faced by media policy, and offer practical solutions.

2. The transition to digital and content regulation: the paradigm shift

Damian Tambini and Stefaan Verhulst

In this first chapter, we aim, first, to outline briefly the so-called paradigm shift in regulation with a focus on 'negative regulation,'[1] (or, regulation to limit circulation of content considered distasteful or harmful). Second, we review the role of self-regulation as an alternative paradigm to government or command-and-control regulation. We conclude by outlining some of the challenges that the transition to digital presents for traditional broadcasting regulation.

Why digital is different

Digitalisation fundamentally changes our communications environment. Analogue broadcasting was characterised by scarcity (of frequencies), few players, one-to-many flow of information, distinctive industry sectors, linear programming, a mediated consumption environment and national boundaries. All this created the shared perception that broadcasting was pervasive in society and that as a result television content needed to be controlled in a structural and restrictive (negative) manner to ensure that public policy goals such as quality, diversity, impartiality, and fairness were met[2]. However, the rise of digital television and the Internet has confounded these traditional understandings of 'television'[3] and broadcasting regulation, and raises the prospect of a new paradigm in communications regulation.

Table 1.1 Regulatory context

The analogue model	*The digital model*
Scarcity and few players (duopoly)	Abundant players (branding)
One-to-many	Many-to-many
Distinctive sectors	Convergence of sectors
Linear programming	Non-linear programming (on-demand)
Mediated consumption environment	Dis-intermediation and individual consumption environment
National boundaries	Transnational and global

In the old world, it was justified to intervene in content markets because there was a clear public expectation of intervention. In the new, digital world, we must have new assumptions about abundance and changes in the mediation process. The information flow and amount of channels available have steadily grown as digitalisation and cheaper content production has made more data publicly available. So-called 'dis-intermediation'[4] has developed a new content producer—user linkage without any indirect barrier or mediation and that is based upon non-linear programming (ranging from on-demand to interactivity). Moreover, it appears that within this new digital setting, some power has shifted from the service or content provider to the user, creating a many-to-many communications environment. The significance of the 'old intermediaries' such as regulators of content is arguably decreasing. Finally, the new digital services are global, recognising no national boundaries. In addressing these new regulatory questions, governments do not have the comfort of being able to consider the issues simply within the confines of their territorial borders.

Why regulate content?

Because convergence blurs the boundaries between the previously distinct technologies of broadcasting and telecommunications, it challenges us to return to the basic justifications of regulating broadcast content.

In the old world, where there was a scarcity of channels, viewed mainly by families or in semi-public settings, it was justified for democratically accountable public institutions to intervene and regulate what was seen. The rationale for intervention hinged on liberal democratic notions of the particular importance of media, and notions of market failure and harm to individuals, particularly children. In the new world those basic philosophical justifications shift, and do so in a radical way.

Regulation of mass broadcasting was seen as justified because of the particular features of broadcast communication, which distinguish it from other media such as telephones and the press.

- *Pervasiveness.* It is difficult to escape the reach of broadcasters. If they are not encrypted, broadcast signals can be played by all receivers within the range of transmitters. Because each transmitter can carry only a limited number of channels, and 'the broadcasting spectrum provides only a limited space for programming, each programme has been very pervasive. There are televisions in more than 95 per cent of all homes, and the five main channels are available across the UK. Surveys show that around 70 per cent of citizens cite TV as their primary source of political information (ITC 1999). In the 1950s, when televisions could only access one channel in the UK, it was imperative to ensure that that channel reflected the opinions, taste and culture of as many different people as possible. In other words, that there was something for everyone. In the 1980s, media policies were developed to ensure that political, ethnic and other minorities had more presence on television. These policies were justified because of the pervasiveness of the medium of TV, and of the range of channels available. The five main channels are received in every TV home. These channels have detailed provisions on what can and cannot be broadcast, which do not apply to less pervasive channels.

- *Invasiveness.* Televisions, radios, and Internet enabled PCs are in our homes and in our children's bedrooms. We often turn on televisions without knowing what is scheduled for that time, and would not like to be confronted by an image we consider offensive or experience as harmful. Content accessed in books, and via the Internet is more likely to be deliberately chosen. When content is invasive central negative content controls should apply (DCMS 1999b). Those forms of delivery where adult users have a high degree of choice and control are subject to less regulation than free to air television.[5]

- *Publicness.* All societies have taboos that are crucial for monitoring social cohesion and individuals' sense of trust and community. Taboos are excluded from 'public' communication and conversely public communication is reserved for topics and content that is considered more valued. The broadcast media are now the key place where society thinks about and discusses the crucial issues about our survival and purpose. The contents they carry constitute a large part of what individuals have in common and provide key reference points in broader political and cultural debates. They are also the stage where we display role models and good examples for society as a whole.

- *Influence.* A radio or TV programme can bring down a government, or make an individual's life very difficult to lead. They can also change culture by defining the norm. Broadcasts can also have a more subtle influence in agenda setting: by determining the focus of public political debate each day. The high degree of influence of broadcast content warrants close public regulation of potential abuses, and makes it imperative to develop the means to ensure that content is not used in a manipulative way. Where there are very few dominant sources of information, or just a handful of television channels, then their influence can be of paramount importance in forming public opinion. It is justified to ensure that such content is neither harmful nor undemocratic, and that it promotes publicly shared values, rather than those of a powerful minority.

Where these features remain, content regulation accountable to the public should continue to apply. But in the changed situation it is simply not possible to carry on as before, as Forgan and Tambini show in Chapter 3. Old instruments, such as detailed command and control regulation of content seem on the one hand ineffective if they can be circumvented on the Internet, and on the other hand costly and damaging to a key industry. At worst, in the new environment the very same policies that seemed justified in a situation of spectrum and channel scarcity seem to be unjustified, unnecessary nannying or even coercive. Competition and consumer choice will play a greater role in the new world, but policymakers must have a clear sense of their limits: what will they not deliver?

How to regulate content

Technological change is dramatically altering the ways in which publicly accountable bodies can regulate media markets. Digitisation, market fragmentation and the Internet have many implications for public broadcasting and positive content regulation. Yet, many point to the Internet and argue that regardless of what governments aim to regulate, there are some forms of content delivery that are patently impossible for them to regulate in the traditional manner. Experience from countries where more proactive Internet regulation has been attempted, such as Australia, bears this out (Tambini 2000b).

The traditional means of regulating through licensing broadcasters and imposing conditions on their licenses are being challenged by the sheer volume of material transmitted. This is of course not true of all means of delivery, and many conventional broadcast services can continue to be regulated as before. The lack of centralised command and control oversight of content does not mean chaos, or anarchy, however. Where central regulation is impossible, but a rationale for regulation remains, there can be a general shift to co-regulation or self-regulation. This refers on one hand to regulation of their consumption by consumers themselves using conditional access, filters or passwords, and on the other to the exercise of restraint and codes of practice by communications companies themselves. It has been generally accepted for some time that self-regulation will play a greater role in the future. However there are some goals that self-regulation will never achieve, and some where the most effective way of policing self-regulation to ensure that it delivers remains to be seen.

Broadcasting regulation was and is based upon a licensing system restricting what kinds of content should be broadcast through guidelines governing the portrayal of harmful content, specific programme schedules or watershed policies and the classification and consequent announcement (acoustic or visual) of programmes. Broadcasters across Europe, as a rule, took their responsibility toward programming content very seriously, and have been keenly aware of the expectations of their audiences, who could complain to statutory bodies in case of offence. Moreover, analogue television was controlled nationally. Finally, access rules and mechanism ranged from ownership rules, must-carry rules, listed events to even public service broadcasting monopolies.

Clearly, these approaches, originally established for terrestrially based mass audience channels, have become much more difficult to maintain as the sources of television programming have multiplied and new technologies have made the prospect of regulation far more unmanageable. Because of converging technologies, it is also increasingly difficult to differentiate between 'telecommunications' and 'broadcasting'. New mechanisms and paradigms appropriate for a changing multi-channel, digital and on-line environment need to be considered.

Taste, decency and safety: the case of negative content regulation

Much of the public concern and debate about new media has focused on illegal and harmful content. This concern has prompted substantial and public industrial response. Negative regulation, in restricting access to content considered harmful, is the element of regulation that faces the toughest challenges in the new environment, because there are now ways of delivering audio and video to the home that are not possible to control in any way. In the search for solutions, various forms of 'self regulation' have been advanced. Four main strands of self-regulation of harmful content in digital services have been suggested and recommended:[6]

- Codes of conduct should be adopted to ensure that Internet content and service providers act in accord with the law and with principles of social responsibility.
- The use of rating/filtering technology and green/red lists should be encouraged.
- Content response and complaints systems must be made available to users, such as hotlines.
- Success requires awareness among users of the means to filter and block content, their power to redress complaints and the level of conduct that is promised by the industry.

It has been argued that the most distinct of these means for enhancing protection and free speech is the use of filtering and blocking mechanisms, using plural and often competing rating systems. This is not to say that legal controls or codes of conduct are unimportant, but rather that the shift from 'hard law to software' can clearly be considered as an empowerment and improvement of users' choice. Hence, the major response to the call for self-regulation involves processes that promote filtering and rating systems.

However, many argue that we are nowhere near the second column yet, and that in the long term many services will remain public, invasive, influential and pervasive. In the meantime, the challenge is to identify the parameters of the shift to digital and work at solutions for the short, medium and longer term.

Table 1.2: Regulatory mechanism

Content control in broadcasting	*Paradigm shift: digital ecology*
Single institutional approach (monopoly)	Plural/competing approaches
Sector specific rules	Content/convey distinction
Media ownership rules (diversity/pluralism)	Access and bottleneck regulation
Government regulation through	Self regulation through
Licensing systems (programme code)	Codes of Conduct
Programme Scheduling (watershed)	Green/Red listing (pool)
Physical bottlenecks (Broadcasting Responsibility)	Technical bottlenecks (Parent Empowerment)
Statutory complaints mechanism	Hotlines

Negative content regulation presents the most concrete case for the possibility for the implementation of a new, self-regulatory regime. There are two main types of development in this field: self regulation of organisations themselves, through agreed codes of conduct, and self regulation by users, through the use of filters/ratings, zonings, and hotlines.

Codes of conduct

Codes of practice or good conduct embody, at times, mutual obligations by competing actors, for example, in the form of agreements, that each of them will take certain action to restrict content that would give any one of them a temporary competitive advantage. Codes provide an indication of the nature of the self-regulatory authority, including whether or not it will impose sanctions for breaches of the code, and upon whom. Codes are often the instrument for the generation or refinement of norms. In the context of the Internet, they are usually the codification of norms negotiated between the industry and government authority. There are a number of industry initiatives underway directed at developing codes of conduct and a number of national governments specifically endorse codes as a front-line mechanism for addressing content issues.[7] However, to be effective a level of compliance needs to be ensured. Users (both industry and consumers) need guidance with regard to the content of these codes and an impartial arbiter needs to have an appropriate redress mechanism, variables often absent with current initiatives.

Rating and filtering

As indicated above, rating and filtering techniques may offer the best solution to empower users/parents in controlling the content they deem to be (un)suitable. Developing filtering and rating systems is always tricky and difficult. Issues of technological literacy, obsolescence of software, international standardisation, and production, distribution or publicisation are potential problem areas.

Zoning

Zoning, or the development of green/red lists, is often considered as a more effective manner in creating a 'safe play ground.'[8] There are, however, several problems with filtering based on lists of sites to be blocked. Due to the immensity of the Internet, green and red lists are always underinclusive and incomplete. Obsolescence of zoning schema is imminent. Furthermore, and most significantly, transparency is at stake: criteria used by list developers must be made an open and accountable process.

Hotlines

Finally, hotlines have been suggested as an appropriate content concern response system for a decentralised and global environment such as the Internet. However, as Burkert has indicated,[9] the quality, rules, procedures and organisational contexts of these hotlines play an important role in the acceptance, functioning and effectiveness of content concern response systems. Hotlines need to establish efficient communications systems, with availability, reliability, and transparency as the main qualitative requirements.

In view of the example of negative content regulation, it appears impossible for auntie, or a paternalistic regulator to vet all video content that is transmitted into the home. Self-regulation will become more important in the process of protecting individuals, especially children. Clearly in the long term this will be a change that warrants the term 'paradigm shift' in communications regulation.

Is self-regulation the future?

The example of taste and decency/safety regulation shows that, in the medium to long term, we are indeed witnessing a fundamental shift in communications and media markets and regulation.[10] Where regulation is still required, for example, to protect children, but existing regulators can no longer deliver, self-regulation can achieve a great deal. As many broadcasting experts are quick to point out however, this shift will not happen overnight. In the short term the core broadcast channels will continue to be pervasive, invasive, public and influential, and consumers expect some degree of regulation.

Currently, the laws that serve as the regulatory framework to control content are coming under increasing scrutiny. More and more limits on the capacity of such laws to police change are suggested as new challenges arise as a result of the digitalisation of content. In response, governments and industry world-wide have been searching for a new regulatory paradigm. The Fourth Report from the House of Commons Culture, Media and Sport Committee, Session 1997-98, entitled *The Multi-Media Revolution*,[11] concluded for the UK that:

> ...the current system, composed as it is of numerous bodies with confusing and overlapping jurisdictions and powers often ill-fitted for digital technologies, is more reminiscent of a feudal State than a regulatory structure for the multi-media age. The case for change is unanswerable (para 157).

We have seen that negative content regulation is one area where self-regulation has been proposed. However, the extent of the viability of self-regulation for other areas of regulatory concern, in positive content regulation, domain-names, standard setting of conditional access systems, open access and e-commerce, remains to be fully explored. While a new consensus is emerging around the role of self- or co-regulation, rather than de-regulation,[12] the debate should be shifted beyond such simplifications and should carefully examine the advantages and disadvantages of implementing a new regulatory settlement.

The professed advantages of self-regulation over governmental regulation include efficiency, increased flexibility, increased incentives for compliance, reduced cost, and minimised government intrusion in the speech field.[13] Potential disadvantages include the danger of regulatory capture by industry and inadequate conception of the overall public interest. As the role and limitations of self-regulation are explored, the following should be taken into account.

First, self-regulation is neither deregulation nor non-regulation. Deregulation directly aims at removing any regulation perceived to be excessive and hindering market forces. Self-regulation, as an ideology or approach to management, does not aim primarily to dismantle or dispense with a framework for private activity.

Second, self-regulation involves a dialogue between government and a business association. The form of the dialogue depends on the situation of each party at the beginning of discourse. This may differ across sectors, across countries and in time. As a result, self-regulation has and will continue to have different meanings in different contexts. Self-regulation rarely exists without some relationship to the state, which is a relationship that varies greatly.

Third, it is naive to suggest, as is done continuously, that self-regulation does not itself involve significant regulatory costs. For any system, regulatory or self-regulatory, costs are determined by a combination of the policy goals they envisage and the structures and dynamics of the economic and social activities they regulate. Monitoring and evaluative costs otherwise assumed by governments may be incurred directly by individual companies or indirectly by industry associations that will also generate significant third-party costs associated with governmental compliance procedures.

Fourth, self-regulation can quickly become moribund without strong and committed support for its development, implementation and enforcement. The very nature of a voluntary system potentially creates a 'free-rider problem', where some actors expend significant resources on the development, monitoring and implementation of codes and standards while others ignore their existence. To be a living and working instrument, a code of conduct or its equivalent must, in practice, be implemented with the agreement of the industry sector to which it applies, and active involvement by government or regulator.

Fifth, effective self-regulation requires active consumer and citizen consultation based upon shared responsibility at all stages of development and implementation. Without user involvement, a self-regulatory mechanism will not accurately reflect user needs, will not be effective in delivering the standards it promotes, and will not create confidence. A general point is that if the public is to bear more responsibility it has to be aware of its role, and we do not yet have the necessary levels of public awareness.

Table 1.3: Self-regulation

Advantages of self-regulation (over governmental regulation)	*Considerations and challenges to self-regulation*
Increased efficiency	Is almost always a misnomer (relationship with government)
Increased flexibility	Is a paradigm different from deregulation or non-regulation
Increased incentives for compliance	Involves significant regulatory costs
Transnational scope	Requires active consumer and citizen consultation (corporatist)
Increased expertise	Needs strong and committed support from all parties (social responsibility vs free-riders) to be effective
Reduced cost	Works best when there is a balance between the carrot and the stick.
Minimised government intrusion in the speech field	

However, it is neither possible nor advisable to regulate on the basis of a notion of an uncertain future. We have no historical example that can inform policymaking in this area. We do not know whether the shift to digital will have the widespread effects on consumer behaviour that some people predict. But there are some things that regulation can facilitate and encourage, and there are barriers that regulation should seek to remove. The challenge is to regulate for the transition.

Summary of principles and challenges

Regulating for the transition: a new settlement

The government must adjust communications regulation to the changes taking place. Despite the best expertise, efforts and traditions of the regulators, the existing archipelago of regulatory authorities is already sinking under the rising waves of technological change. Reform is necessary not only to protect the public interest based upon the traditional rationales, such as protecting children from harmful content and meeting public expectations in provision of quality content. It is necessary in order to permit and encourage the development of a national communications infrastructure that will determine the competitiveness of UK plc. The overall structure of communications regulation should be rendered more efficient, proportional, flexible and responsive and regulation should be undertaken only when necessary. In the current period, the rate of technological change constitutes in itself a challenge to regulatory institutions with those goals.

New digital technologies do transform the justifications, context and tools of media and communications regulation. Many of these arguments apply also to positive content regulation and the necessity of ensuring a democratic and plural media. But many of the traditional goals and objectives of regulation remain intact, since it remains as important as it ever was to have a media environment that is as safe, fair, plural and enlightening. The European Green Paper on Convergence partially dissents: 'The fundamental objectives underpinning regulation in the Member States are not undermined by convergence' (European Commission, 1997). 'Nevertheless,' it continues, 'the nature and characteristics of convergence as well as the perceived need of industry actors for regulatory intervention to be limited and closely targeted, should lead public authorities at both a national and a European level to re-examine the role and weight of regulation in a converging marketplace'.

In the long term there will be a greater role for self-regulation and co-regulation, particularly in relation to negative content regulation. This is not a negation of responsibility by government but involves significant government engagement. There remain transitional difficulties and there are drawbacks and uncertainties about where self-regulation will deliver and where it will fail. The paradigm shift will not, however, arrive at Internet speed. The traditional tools of content regulation for broadcasters – licensing conditions – will not and should not disappear overnight, and there is no guarantee that the traditional justifications for regulating content will disappear either because many channels of delivery will remain pervasive, invasive, public and influential.

Clearly there is controversy regarding just how much the transition to digital will impact on broadcasting and its regulation; how quickly it should do so and what the public interest is in the new environment. Luckily, policymakers do not need to know exactly what the future holds, but can regulate for a range of possible outcomes. The contributions to this volume reflect the following general principles:

- The starting point of every regulatory intervention, including content regulation, must be the policy objective, not the means of achievement.
- There is a role for government to foster and encourage, not merely respond to, the shift to digital and Internet access.
- Communications reform should, where possible, maintain standards of serving the public interest in content regulation. Public interest objectives do not become irrelevant or invalid as a result of technological change.
- There are new aspects of the public interest that should inform policy in the new environment, which cannot be inferred simply by combining public interest principles from telecommunications and broadcasting: Human Rights Act principles, which underline communications entitlements and freedoms, should be observed.
- Finally, a functional approach is needed, one which does not depend solely on technology or forms of delivery, but which recognises the nature of the service and the character of the audience receiving it.

A settlement for the transition: a continuing role for the regulator(s)

Although in the long term there will be many ways of delivering video and audio content to the home whose regulation will be neither possible nor justifiable, public regulatory institutions still have many cards to play in the foreseeable future:

- First they still control spectrum privileges. Broadband cable, satellite and other platforms promise in the long term to devalue government control of market entry as a tool of content regulation, but it does have purchase in the meantime, because cheap or free spectrum can be provided to services that conform to publicly accountable standards. Those who receive those benefits can be expected to deliver public service.

- EPG prominence will determine audience for some services, and can be afforded to those services that fulfil public service obligations.
- Must-carry (and must provide) rules will increase the pervasiveness, invasiveness, publicness and influence of channels. Those that benefit in this way should be subject to corresponding public interest content regulation.
- Public funding is still necessary to promote educational, informative and cultural content that the market would not otherwise provide.
- And of course with extreme cases, general law on libel, child porn and defamation does apply in the online environment.
- Government can engage with industry in development and policing of codes of conduct and compliance, and in promoting greater public awareness of user's responsibilities for promoting safe use of the media.

Communications reform must start from the basic assumption that the shift to digital is itself in the interest of the public. There is a clear public interest in the rapid provision of open, any to any communications infrastructure to foster economic development, access to culture and education and democratic communication. The government should put into place a system of regulation that will foster and encourage the digital transition, rather than obstruct it. In practice this means a streamlined set of institutions with clear sanctions and well-defined goals, to replace the current fragmented structures. Regulatory reform also requires active consumer and citizen consultation based upon shared responsibility at all stages of development and implementation. Without involvement of the public voice, a regulatory mechanism will not accurately reflect user needs, will not be effective in delivering the standards it promotes, and will fail to create confidence.

Policy must also protect the delicate ecology of broadcasting. The recommendations listed at the end of this book outline what government should do to foster the right transition to digital, for there are a variety of potential outcomes of the process. The most dangerous of these is that the transition to digital will debilitate many of the institutions that currently protect the public interest, without the emergence of new institutions that help to bring out the cultural, democratic and educational potential of the new media.

Endnotes

1 See Waverman L, *Broadcasting Policy Hits the Internet.* Available at http://www.gip.org/gipwp40500.htm

2 These public policy goals are generally based upon the view that broadcasting is a means of communicative self-development and important to the functioning of a democratic society. For a detailed analysis of the rationales behind broadcasting regulation see Hoffman-Riem W, *Regulating Media. The licensing and supervision of broadcasting in six countries* New York: The Guildford Press, 1996, p297

3 See Owen B, *The Internet Challenge to Television* Cambridge, Massachusetts: Harvard University Press, 1999

4 Or the removal of intermediaries from the industry value chain. See Moore C, *Disintermediation: Communications Technologies are having some impact* in Dataquest Interactive: Ask the Analyst, 19 March 1996

5 *Regulating Communications The Way Ahead* Results of Consultation Paper, Department of Culture, Media and Sport 1999, at 1.14.

6 The European Internet Action Plan lists these three ways as crucial to create a 'safer' Internet environment. See http://www.ispo.cec.be/

7 See Blinderman E, Price M and Verhulst S, *Codes of Conduct and Other Self-Regulatory Documents: Emerging Patterns of Norm Formulation and Enforcement on the Internet* Bertelsmann Foundation, 2000

8 See for instance the conclusions of the Which? Study on rating and filtering techniques. Available on Which? website: http://www.which.org.uk/

9 See Burkert H, 'The Issue of Hotlines', in Waltermann J and Machill M (eds), *Protecting Our Children on the Internet: Towards a New Culture of Responsibility* (Bertelsmann Foundation Publishers) (2000)

10 See for instance Verhulst S and Marsden C, *Convergence in Digital Television Regulation* Blackstone Press, 1999, Verhulst S, Goldberg D

and Prosser T, *Regulating the Changing Media* Clarendon Press, 1998 and also COM (1999)108 EN *Final Results of the Public Consultation on the Green Paper* http://www.ispo.cec.be/convergencegp

11 Select Committee On Culture, Media And Sport Fourth Report, The Multi-Media Revolution – Volume 1, *The Multi-Media Revolution* (1998) 21 May 1998, HC 520

12 See also Prosser T, *Law and the Regulators* Oxford: Clarendon Press, 1997 – Chapter 10

13 See Price M and Verhulst S, 'The Concept of Self Regulation and the Internet', in Waltermann J and Machill M (eds), *Protecting Our Children on the Internet: Towards a New Culture of Responsibility* (Bertelsmann Foundation Publishers) (2000)

3. Through with ownership rules? Media pluralism in the transition to digital

Damian Tambini

The UK, like other countries that wish to remain democratic, has laws limiting the number of media companies any one individual or company can own.

The rules in fact have two aims. First they seek to promote effective competition in media markets by placing structural limits on market share. Second, they seek to promote the dissemination of a plurality of independent and accurate information and views, which an unregulated market would fail to provide. In terms of this latter objective, ownership rules are a crucial defence against attacks on some of the less robust aspects of our democratic constitution. They protect against the eventuality that one single company or individual could have excessive control over public opinion or the cultural zeitgeist.

With changes in media markets, many have argued that the existing rules are no longer effective, that they hinder business, and that they can be phased out. Others reply that any relaxation of ownership rules would be a fundamental threat to our democracies because it could open the way for concentration of ownership and intensify content controls in the hands of owners. The resulting standoff is creating a policy logjam that is not only unnecessary, it is potentially damaging to the UK economy.

In order to judge whether the time has come to reform ownership rules, we need to return to their fundamental objectives: why do we have rules to prevent concentration of media ownership, and are those reasons still valid in a rapidly changing environment with the shift to digital?

This chapter reviews the basic rationale of media ownership rules and proposes a strategy for protecting and promoting media pluralism in the transition to digital. It argues that although it is highly likely that the rules could be relaxed in the long-term future without too much damage to democratic communication and pluralism, it is too soon for significant reform of the rules in broadcasting markets. The 'convergence rationale' for relaxing the ownership rules has been exaggerated: whilst competition regulation alone might be a means of guaranteeing adequate pluralism in the future, it is not yet, and even in new media markets such as online and the internet, worries about pluralism remain. We should phase in new competition rules that will guarantee pluralism in digital broadcasting, and hold firm on existing rules for analogue.

The high level of merger activity in communications industries reflects the ongoing processes of globalisation and convergence that are transforming the sector. In the current context, significant economies of scale and scope can best be realised through common ownership of different products and enterprises, and integration along the value chain. Where a company controls both networks and content there is a threat not only to competition but to plurality if that company seeks to deny competitors' content access to its networks. The increasing role and place of the media and information sector in generating wealth for the UK presents a formidable challenge to the formulation and implementation of practical policies for achieving diversity and impartiality.

During the transition to digital, democracy and pluralism must come first, even before economic growth. The central challenge is to devise a policy strategy that can realistically protect and enhance media pluralism during the transitional period from analogue to digital. This strategy should combine old with new approaches to competition, and also introduce new measures to democratise the media.

Historical background

Since the 1947 Royal Commission on the press first tackled the issue of concentration and ownership policies for the newspaper industry, a consensus on media ownership has more or less held firm across the political spectrum. Rights of freedom of expression and accuracy of news presentation (Feintuck, 1998: 94) together with the need for a plurality of sources of political and cultural information has justified regulation to prevent excessive concentration of media ownership. The primary concern of these rules was to defend democracy: this is why the limits on media ownership were set at levels more stringent than that demanded by normal competition rules in other markets. (Beard, Cave and Holmes, 2000). This has applied not only to newspapers, but to broadcasting, and cross-media rules have specified limits on ownership across press and broadcasting.

In 1962, the Pilkington Committee noted that cross media ownership threatened democracy since excessive concentration of power might be used to persuade and influence public opinion and to present a one sided presentation of public affairs. Their conclusions resulted in the development of a power to bar newspapers from holding shares in programming companies when this share goes against the public interest (Pilkington, 1962; Gibbons, 1998: 220).

Ownership rules for television and radio appear in a variety of forms. The Fair Trading Act 1974 places limitations on newspaper ownership. The Broadcasting Acts of 1990 and 1996 specify clear limits on ownership and concentration. There are detailed rules governing cross media ownership that place specific limits on the degree to which individuals or companies should control interests in combinations of press, radio and television sectors.

With some exceptions, public interest tests are applied in cases where a relevant broadcasting licence holder becomes connected with a newspaper or in large broadcasting mergers. When deciding if such a merger is sufficiently consistent with the public interest, criteria applied include the potential impact on diversity of public information and opinions expressed on television or radio or in newspapers.

The Broadcasting Act 1996 updated existing broadcasting legislation to include digital broadcasting, amending the 1990 act, while maintaining the existing competition legislation. In introducing the legislation, Virginia Bottomley, then National Heritage Secretary said that:

> change needs to be embraced but – in a democracy – no one organisation should be able to dominate all our media. Through the Act we have succeeded in achieving a balance between proper commercial demands and the wider public interest which requires plurality, diversity of opinion and the highest possible standards.

The ownership and concentration rules, as Richard Collins and Cristina Murroni put it, thus 'kill two birds with one stone', providing structural limitations on market power, and at the same time promoting media pluralism. Questions of reform and updating of ownership and concentration rules must therefore be aware of their dual nature (Collins and Murroni, 1996).

Ownership regulation has been discussed increasingly often at the level of the European Union over the last decade. Media pluralism and concentration have been topics of discussion at the Commission, Council and Economic and Social Committee. As early as 1990 the Parliament published a resolution pointing out that 'restrictions on concentration are essential in the media sector, not only for economic reasons, but also and above all as a means of guaranteeing a variety of sources of information and freedom of the press.' (Cited Levy, 1999, 51). The Commission in a 1996 proposal suggested a 30 per cent upper limit on monomedia ownership for radio and television broadcasters in their own transmission areas, and an upper limit for total media ownership – ie ownership of television, radio and/or newspapers – of 10 per cent. Modified proposals were released in the 1997 draft directive, *Media Ownership in the Internal Market*. The lack of Commission progress on this issue (Commissioner Monti rejected the proposals after pressure from some member states) reflected the fact that the Commission's competencies are stronger for competition law than they are for content. (Doyle, 1997). Also, and more fundamentally, the central *democratic* concern of ownership rules has reflected democracy on the nation-state level. National governments resist application of pluralism principles at the sub-national or European level.

The history of ownership rules shows that they have been defined in negative terms; by identifying the dangers to be avoided rather than positive goods to be promoted. Restrictions on ownership are a safeguard against any organisation or individual hijacking democratic debate. This negative definition hinders reform, since it results in a contested notion of what pluralism is, which is difficult to translate into concrete policy instruments. If we are to develop a progressive approach in this field it is imperative to develop a clear notion of what it is about concentration of media ownership that worries us. Only then is it possible to ask if there are tools other than ownership restrictions that could achieve those same objectives.

What is media pluralism, and why does it matter?

The wooden language and cultural stultification of communist societies amply illustrate the dangers of concentrating power over opinion formation in too few hands. Though of a lesser degree, similar dangers can appear in capitalist media markets.

Democracies are founded on principles of freedom of movement and opinion, and of political and civic equality between citizens. Given the clear influence that media reporting and presentation can have on public opinion, the prospect of a single player controlling the 'marketplace for ideas' is contrary to the public interest. The dangers of a failure of pluralism are clear: As Andrew Graham and Gavyn Davies put it: 'votes should not be bought and sold' (Graham and Davies, 1997 p31).

Those who seek firm positive definitions or measures of media pluralism, however, are likely to be disappointed. Ownership rules do not actively promote pluralism, diversity and the free exchange of ideas. Rather they seek to protect these goods from a variety of threats that result from the natural tendencies, in media as in other markets, towards concentration of ownership and integration along the value chain. At the same time as industrial policy concerns create pressure to relax limitations on ownership, current technological and market changes are intensifying concentration and vertical integration, but they are also changing the relationships between content producers, networks and consumers. New concerns emerge in terms of competition: there is a danger that a firm with market power which owns both the network and content will squeeze out competing content. These trends have implications not only for competition, but also for pluralism. It is not the concentration of ownership *per se* that we are worried about, it is the effect it has on the broader democracy and freedom of opinion.

It is the dual objective of ownership rules that have created problems in their reform. Whilst their object and limitations with reference to competition law can be clearly calculated (see Beard, Cave and Holmes, 2000), their performance in protecting pluralism is less easy to define, precisely because of the lack of a positive definition of pluralism.

According to the Council of Europe Committee of Experts, the benefits of media pluralism are:

> To serve democracy by providing the citizens with a broad range of information and views needed for the effective exercise of citizenship; to give minorities the opportunity to maintain their separate existence in a larger society; to limit potential social conflict by increasing the chance of understanding between potentially opposed groups and interests; to add to the richness and variety of cultural and social life; to open the way for cultural and social change.[1]

For an idea to be effective in policy and regulation it must be clearly defined. Media pluralism is not, and this is a central challenge. Notions of pluralism, diversity and the marketplace for ideas are at best vague and malleable, at worst adjusted to the purpose of whoever invokes them. Nonetheless it is instructive to examine these attempts at positive definition, because they help reveal what is really at stake.

A search of recent research literature on pluralism, diversity and media ownership yields dozens of separate research studies that have in recent years attempted to define media pluralism and diversity (See Iosefides, 1997; 1999: 154-5; Napoli 1995; Napoli, 1996). The growing trans-national literature reveals a range of competing definitions of media pluralism and diversity as the objects of ownership regulation. Some documents refer to the necessity of promoting 'freedom of opinion' and 'freedom and variety of expression of opinion' (Department of Trade and Industry, 1999). Others refer to *the free marketplace for ideas* or draw a distinction between *source, content and exposure pluralism or diversity*. Whilst policy is a messy game, some clarification of these ideas and their role is necessary if the optimal solution is to be found for the transition to digital.

The *marketplace for ideas* metaphor has guided the debate in US courts since at least as long ago 1919 when Oliver Wendell Holmes stated that 'the ultimate good is better reached by free trade in ideas. That at any rate is the theory of our constitution.' In the US, the idea of the marketplace of ideas has been used by the FCC since the 1960s. Recently, the 'marketplace ideas' metaphor has been interpreted in a quite literal way in the US debate, with the notion that ideas are traded in a competitive marketplace and the better ideas generally gain market share. There are of course many assumptions inherent in this view, not least that perfect competition, genuine plurality, ease of market entry and perfect information exist in this marketplace for ideas.

Diversity, particularly in the US context, often includes the ethnic, regional and gender breakdown of communications professionals and content creators. The FCC, for example reported in 1998 on a study of gender diversity of licensees, in order to 'prove the link' between diversity of sources of information and diversity of content. (Kennard 1998; Napoli, 1999 7-8).

The notion of *pluralism* is used more often in Europe. Jay Blumler's (1992) notion of *pluralism* focuses on (i) plurality of content, (ii) access to different points of view, and (iii) geographical representation.

Source pluralism, content pluralism and exposure pluralism cross cut with notions – more familiar in the UK; of *internal and external pluralism and diversity*, which refer to the differing policy goals of ensuring content or source pluralism within or between texts/ channels. For the purposes of clarity in this paper we will use a distinction between source, content and exposure pluralism (Napoli, 1995), and dispense with the confusing distinction between pluralism and diversity.

In an ideal world, citizens are free to choose from a wide range of ideas (content pluralism) delivered from a wide range of sources (source pluralism). The citizens then partake of this pluralism (exposure pluralism) to increase their knowledge, encounter opposing viewpoints, and become well-informed decision makers who are better capable of fulfilling their democratic responsibilities in a self-governing society (Napoli 1999, 9). Where a few media owners control information this is not possible, and democratic citizenship is under threat.

Table 2.1 outlines some of the various ways in which media pluralism has been defined.

Clearly, ownership rules are not the only policies that seek to guarantee a plural and diverse media. They need in particular to be seen alongside positive content regulation, and the broad tradition of public broadcasting. By specifying a balanced and impartial approach to 'matters of political and industrial controversy' the regulation of news guarantees 'internal pluralism'. Public broadcasters such as Channel Four in the UK have the specific remit of representing diversity of views and lifestyle.

To guarantee source pluralism is not necessarily to guarantee content pluralism. (A variety of sources may put out similar programme formats and viewpoints, and all may neglect ethnic minorities.) Equally, the mere

Table 2.1 Three aspects of media pluralism

	Source pluralism	*Content pluralism*	*Exposure pluralism*
Policy goals eg	Plurality of sources of info and editorial in market as a whole Limit audience share of any one owner	Plurality of programme-type format in market as a whole Plurality of demographic group represented • Ethnic • Regional/national • Political viewpoint	Plurality of content and sources exposed to each individual. Within channels/ texts (internal) Must-carry/positive content between channels/ texts (external)
Policy instruments eg	Audience/advertising share limits, Public Service Broadcasting, public interest tests for mergers	PSB, positive content requirements (eg impartiality) programming quotas, public interest tests for mergers, audience/ advertising limits	Must carry PSB, cross media ownership rules, public interest tests for mergers.

After Napoli, 1999, 10; cf McQuail 1992

provision of a variety of types and sources of content does not exclude the possibility that the vast majority of a given audience might be exposed to a particular source and type of content. This is why many have argued that there should be a broad and ambitious notion of pluralism that ensures that both internally plural and externally plural services are delivered.

A positive definition of pluralism is therefore difficult to agree. Indeed, because ownership rules have always acted as a proxy for the goal of media pluralism, we never had to have a positive definition of pluralism. It is much easier to provide a negative definition: to isolate the phenomena that ownership rules seek to avoid. The following are examples of dangers where pluralism is compromised:

- *Source pluralism*
    - Company interests put before public interest
    - Excessive individual power to form public opinion or pursue a political agenda

  - Sidelining or by-passing of the democratic representation process

- *Content pluralism*
  - Ethnic/ regional exclusion from public representation
  - Excessive individual power over opinion formation, political agenda setting
  - Cultural stagnation or homogenisation due to lack of choice or 'Hotellings effect'

- *Exposure pluralism*
  - Restrictions on individual freedom
  - Narrowing of knowledge available to individual
  - Narrowing of cultural choice

At the most general level, the worst case scenario to avoid is when an individual or company can leverage power in communication markets into political power, in a self-reinforcing cycle. At the policy level, the question is which interventions can best prevent such abuses, given the broad technolgical market changes currently being experienced.

Two main principles of media pluralism must be reaffirmed:

- No single company or interest should have control information distributed in society as a whole. (source, content). This guards against political and social hijacking of democratic process and cultural change.

- No single company or interest should control information distributed to each individual (exposure). This guards the fundamentals of liberal freedom: freedom to form opinions, freedom of thought.

Competition policy and media pluralism

Clearly, pluralism matters. It has long been accepted that an effective competition policy, whilst necessary, is not sufficient to create the media landscape required by democracy. What is at stake is whether digital makes any difference and what form of policy and regulation can guarantee pluralism in the digital environment. In this section I outline the role of competition in ensuring plural media markets, and in the following I deal with the issue of pluralism in the transition to digital.

The goals of the maintenance of pluralism and competition are at times, although not consistently, convergent. In cases where goals are divergent, different approaches need to be applied (Aitman, 1994). Competition policy seeks to prevent anti-competitive behaviour, that is, where a producer raises price above cost for a sustained period. It is based on economic theory and evidence suggests competition between many suppliers will drive price down towards cost while promoting innovation and quality (Geroski, 1991). Where substitutes for particular goods or services exist, suppliers will find it difficult to market products inferior to or more costly than substitutes. Where there are no substitutes for the good, the producer holds a monopoly and is potentially able to earn supernormal profits. To prevent this, competition policy seeks to encourage the provision of substitutes by removing barriers to entry into particular markets (see Beard, Cave and Holmes: 3, 4).

Competition policy promises to define and measure abuses in the marketplace for goods and offer remedies. Moreover, it promises to promote source pluralism (many providers) and also differentiation between the products and services provided. The theoretical potential of competitive markets to respond to demand signals from consumers and to engender pluralism and quality without political intervention in markets explains why the application of competition policy principles to the marketplace for ideas has seemed so attractive.

There are several reasons why even perfectly competitive media markets will not guarantee adequate pluralism in the media. First, competition policy might tolerate a market with a restricted number of providers, or indeed a single provider provided that this dominant player did not engage in anti-competitive behaviours. Economic efficiency might result in service from a huge, efficient provider. If there were no barriers to entry and the market in question was deemed to be contestable (Baumol and Sidak, 1994), a regulatory regime based in competition policy might find this unexceptionable. The relevant market might be defined as including companies not currently making the product or offering the service in question but which could switch to doing so (London Economics, 1994, Dertouzos and Trautman, 1990). Such a 'winner takes all' market might (or might not) be competitive but would not be characterised by pluralism of source, content or exposure.

Second, many doubt that competitive media markets, even those with more than one firm supplying relevant products and services, do deliver diverse, quality media products. They characteristically point to broadcast television in the US where there is significant convergence of programming content. Blumler found that the US terrestrial television regime was 'inimical to broadcasting range – of programme form, of sorts of quality favoured, of viewer affect and experience stimulated' (Blumler 1986: 141). The simultaneous screening of the OJ Simpson trial on seven separate channels in the United States has done little to challenge Blumler's judgement.

This diagnosis can be likened to what economists name the Hotelling effect. Hotelling proposed that economically rational competitors will always tend to crowd in the middle of the spectrum of consumer tastes rather than provide a diverse range of products (Hotelling, 1929). Although valid where there are three or four suppliers (for example, broadcast channels), this claim may be less convincing when viewed in context of media abundance, where there are ten competitors versus three. There has to be a critical point after which it is more rational to pitch products and services to a market niche than to intensify already hot competition in the middle-of-the-road. Thus, Hotelling's analysis helps explain the evolution of parallel schedules and programmes of BBC1 and ITV, but may be less useful in a world of tens, if not hundreds, of media channels, competing on price as much as on quality. Applied to new media, Hotelling's effect suggests that the middle ground of each niche market may be crowded, but does not imply proliferation of hundreds of identical general entertainment channels with the explosion of bandwidth. The theory does, however, reinforce the recognition that a limited number of suppliers in a particular market may be inimical to pluralism. There may also be effects less easily written into the assumptions of the economists, such as a tendency for individuals to value certain 'common knowledge' – shared facts and cultural practices – simply because they are shared.

In this situation, the implications for ownership are not straightforward. It is economically rational for a single organisation, or interest, to spread risk and not compete against itself by marketing products in several different markets or market sectors rather than in one. Just as individual investors may spread risk by investing in a unit trust or investment trust rather than in a single company so too is it rational for companies and owners of media corporations to do the same. Moreover, if an owner has a company competing in one media market sector it will usually make better sense to position sister companies in different market sectors with a consequential improvement in the overall risk/reward ratio. Cross ownership can therefore, at least in theory, support rather than inhibit diversity and pluralism. Here too there are implications for public policy which competition policy is unlikely to recognise. If each separate market in which a single firm owns one supplier is functioning well then no grounds for intervention arise from considerations of competition policy. Grounds for intervention might, however, exist if the media were regarded as a single market, rather than separate markets, or on the separate grounds of pluralism.

Third, whereas competition policy emphasises substitutability and therefore homogeneity, this conflicts with the aim of democratic pluralism. Media products are inherently difficult to substitute for each other. For a true Robbie Williams fan, The Backstreet Boys are unlikely to be a satisfactory substitute. And for a Charlton supporter a sports channel showing Manchester United's games is likely to be of interest only twice a year. The essence of media markets is the attempt to develop unique products for which there can be no satisfactory substitute. True, there is some degree of substitutability in all the markets from which the examples given above are drawn. But, to what extent are The Sun and The Financial Times mutually interchangeable? The lesson to be drawn from the most recent experiment in testing the limits of UK national newspaper markets is that the national daily market is better considered as made up of distinct markets rather than being a single market.

In addition, weaknesses of existing ownership rules persist, including the fact that many facets of control of a media company (such as, direct links, family connections, external financial or contractual agreements) are not adequately covered by ownership rules (Iosephides 1997: 750). This may suggest that the definitions of control in the Broadcasting Act 1996 may not be adequate to the task of ensuring genuine pluralism and diversity.

To summarise, competition policy aims to prevent dominance of communications markets by preventing market dominance and abuses of market dominance. In the past, this was deemed inadequate to promote adequate pluralism in media markets, and specific rules to promote pluralism were devised. Many argue that these – particularly those restricting size and ownership of media corporations – hinder innovation, stunt the development of communications industries, and are becoming increasingly outmoded in a new communications environment. Are ownership rules necessary in digital broadcasting? And what other measures should we consider?

Media pluralism in the transition to digital

A digital dream is gripping the minds of media broadcasters and policymakers throughout the world. It is a dream of an infinitely plural media environment, in which the consumer is in complete control, and problems of pluralism, diversity and media control have vanished.

The digital dream is unfortunately far from reality. Should such a view justify rash removal of ownership rules, the result could easily be a nightmare of vertical integration coupled with concentration constituting a genuine challenge to democratic pluralism. It is precisely the abundance of new digital media that creates demand for new intermediaries and aggregators of content who occupy the position of powerful media gatekeepers.[2] It is precisely the transition to digital that spurs the development of new vertically integrated media companies. Even where channel choice is huge, as in the US, consumers tend still to get the majority of their information from one or several sources. And the appearance of freedom on the net can be deceptive: search engines, commercially driven linking and searching and preferred placement strategies in fact mean that new digital media will provide at least as many challenges to media pluralism as did the old media. Many consumers will stay with the core broadcasting channels, and it is the overall share that we are concerned with, rather than the means of delivery. Even if the digital dream describes the future, there is no guarantee of how quickly we will get there. The challenge in this context is to devise an approach to pluralism that will protect it in the transition to digital.

Current claims about a communications revolution are similar to those made in the 1980s when it was argued that the advent of new cable technologies and market fragmentation would remove the justification of media ownership rules. The promised explosion of TV channels and services did not occur, and the vast majority of our citizens still rely on the mass linear media. (see Humphreys and Lang, 1998: 18-19). Policy should not act prematurely: removing restrictions on ownership under current market conditions without proper safeguards for pluralism could open the door to an intensification of media concentration that would have detrimental consequences for democratic communication in the UK.

Broadcasting *is* changing. It is no longer dominated entirely by vertically-integrated producer-broadcasters, organised around channels that are defined by spectrum position and tuning devices. In the digital environment, the number of channels proliferates and consumer control appears to be greater, but this does not mean that pluralism will be naturally guaranteed by the competitive market. A sophisticated approach to competition that takes into account the aim of pluralism will be the key to the new regime. If new measures can be put in place to guarantee that consumers have access through gateways and networks to genuine choice, then we may well be through with ownership rules in the not too distant future. Measures should be put in place to ensure that competition policy does deliver pluralism in the media sector. The challenge to competition policy is a great one, however, which will require a regulator that is both strong and quick.

Across all sectors, and particularly in those information society sectors where innovation and investment is seen as a priority for the overall competitiveness of the economy, there is a general trend away from structural competition rules limiting market dominance to behavioural rules that focus on abuse of market dominance. This intensifies pressures to justify ownership rules, and to look for alternative ways to promote pluralism of source, content and exposure.

Competition policy focuses on the following potential sites of market foreclosure:

- Conditional access system (STB) that regulates the access of consumers to pay-TV services.
- Conditional access system's implications for access of content providers to audience;
- Monopoly ownership of the distribution system.
- Monopoly ownership of content rights.

Such measures in the transition to digital will not completely resolve our concerns with pluralism. Much ink has been spilt describing how competition policy can respond to the fast changing, innovative world of communications technologies. As Karel Van Miert described in 1997: 'facing gigantic alliances between alternative or complementary networks, on the one hand, and content producers and packagers of information and the carriage networks, on the other, the challenge is identified as balancing the positive synergies of market integration and partnerships against the risks of entry barriers and foreclosure of the markets'. Less attention has been given to the nature of pluralism in those markets, and it has been assumed that pluralism and diversity will naturally be taken care of by competition policy. A council of Europe Report (Marsden 1999) outlined the main potential tools for safeguarding pluralism in the transition to digital. These include:

- Must-carry rules for example for public service broadcasting;
- Gateway and bottleneck scrutiny. (EPGs, API). There may be a threat to pluralism if this is not adequately carried out;
- Open Access. (Guaranteeing access of competing content providers to networks).
- Unbundling exclusive rights or enforcing third party access to CAS networks.

According to Chris Marsden:

> The removal of spectrum constraint offers increased commercial and public service possibilities. Should both result, an increase in pluralism may be predicted. However this will not occur without regulatory design and intervention with attention given to consumer access to diverse programming in an encrypted 'user pays' environment increasingly substituting for ownership and content controls.' Marsden goes on: 'while it would be premature to suggest any deregulation in analogue pluralism controls, it will be necessary in the digital environment to replace structural with behavioural regulation. (Marsden 1999, p12)

Media concentration is likely to intensify in the coming decade. The process of digitisation and convergence has itself acted as a spur to media concentration. According to Camesasca:

> the underlying issues are related to the effect of technological changes on economies of scope...convergence changes market structures. Existing firms in one of the converging markets, in order to compete effectively, seek to enter the other markets, through de nuovo investment or through mergers. These developments are to a large extent escaping existing regulation at the Member State level concerning media concentration (Camesasca, 2000: 85).

Improving pluralism policies

Given the continuing importance of media pluralism, the disadvantages of current policy instruments, and the challenges of digitisation, the following have been suggested as proposals for reform.

- Improve existing methods of measurement
- Promote internal pluralism through public service
- Independent production quotas

Methods

The last Broadcasting Act introduced the 'market share' framework. One response to calls for reform has been to suggest improvements to existing rules, and explore the range of different approaches exists in different European countries. The following have been suggested as ways to improve on our current situation:

- Audience based measures, for example the 'time use analysis' used in 1993 by Arthur Andersen (1994). The basic measure in this case was the 'time the public actually spends in the consumption of media'.
- Share of equity capital as in France and Germany.
- Exchange rates as a means to measure how much value consumers put on each medium.

- Alternative measures such as audience reach.
- Levels of ownership rules should be relaxed: 20 per cent of market share rather than 15 per cent.
- Points schemes, such as the radio points scheme could be extended.

Cavallin (1999) reports that a number of countries currently have quantitative restrictions on media control based on:

- Number of channels (Italy, Portugal, Spain, Sweden);
- Share of audience (UK Germany, France);
- Share of circulation or absolute circulation in the press sector (France, UK);
- Share of foreign holdings (Poland);
- Shares held in one channel (Norway); or
- Right to vote (Sweden).

Clearly measures could be improved. It might also be asked whether there may be instruments other than ownership rules to realise the goal of pluralism. What is clear is that fudging the issue and bringing in minor changes to methodology is not likely to be worth the considerable effort and burden on industry. It is likely to bring in a more unfamiliar and complex system, at a time when technological and market changes could render it redundant in a few years. A key aim of current reforms should be to avoid changes that may have to be reformed in the next 10 years. Tinkering with methodologies is therefore not recommended.

Internal pluralism

Internal pluralism, eg public service channels regulated to provide many viewpoints and groups, can to a certain extent substitute for external pluralism. But in a truly plural market there would be the possibility of accessing both internally plural (for example, PSB) and externally plural services.

In the long term the culture of broadcasting impartiality might be undermined. If there is continued market fragmentation then the ability of regulators and the justification of regulators scrutinising content in any detailed way declines. This does not, particularly in the short term, apply to core linear mass services. In such a scenario the availability of PSB to every consumer is crucial. Internal pluralism and public service can be an excellent bulwark against any tendency for further concentration to threaten public opinion. It is not, alone, however sufficient, and the ideal scenario would be the continuation of a mixed ecology.

Independent production quotas

Independent production quotas, and regional and European quotas help to ensure that a greater degree of plurality and diversity than the market would provide is provided. They go some way to addressing concerns about regional, ethnic and cultural diversity, though not news, current affairs and impartiality. Many have questioned the degree to which regulation will be able to impose production quotas in the new digital environment, so their future as guarantor of pluralism and diversity may be limited.

A strengthened competition regime delivered by a single minded regulator together with the measures described here could deliver an improved competition and pluralism regime. However, another element is required in order to future proof the pluralism regime, reforms which are described in the next section.

Media democratisation reforms

Media democratisation reforms attempt to break the link between ownership and editorial control. These are the most long-term, future proof and practical of all suggestions for guaranteeing media pluralism in this time of change, hype and more change.

The aim of law and regulation limiting concentration of ownership of the media, beyond general considerations of competition policy, is to reduce any possible danger of harm to freedom of expression, pluralism of opinion and accurate reporting of events. It is clear that none of these goals can be achieved through regulation of the ownership of the media alone. For they concern the behaviour of media workers, on whom the quality and character of media content depends, and the behaviour of journalists and editors cannot be 'read off' from the structure and ownership of firms. Professional values enshrined in jealously guarded traditions of investigative reporting, iconoclastic judgement and editorial independence are necessary, if not sufficient, conditions of a healthy media and a healthy democratic polity. How can they be strengthened?

Surprisingly little attention has been given to means other than ownership rules of improving pluralism. Certainly little has been written about journalistic and editorial independence since the Royal Commission on the Press' chapter (interestingly the shortest of the Commission's report) on Editorial Contracts (McGregor, 1977: 154-156). Given the Commission's rhetorical emphasis (albeit an emphasis belied by the brevity of its reflections on the matter) on the relationship between proprietors and editors subsequent neglect of the matter is particularly surprising. The Royal Commission found:

> the importance of publicly and explicitly guaranteeing editorial independence, in the context of concentration, chain ownership and monopoly...is so great that publishers and editors should urgently consider ways of setting out and guaranteeing formally editorial independence from employers (McGregor, 1977: 155).

Democratising media organisations and introducing more democratic decision-making procedures within media organisations would make citizens less nervous about mergers (Keane, 1992; Humphreys and Lang, 1998).

The Royal Commission of 1977 can serve as a basis for defining the aim of media pluralism. It defined the basic rights of editorial and journalistic independence as:

- the right to reject material provided by the central management or editorial services;
- the right to determine the contents of the paper (within the bounds of reasonable economic consideration and the established policy of publication);
- the right to allocate expenditure within a budget;

- the right to carry out investigative journalism;
- the right to reject advice on editorial policy;
- the right to criticise the paper's own group or other parts of the same corporate organisation;
- the right to change the alignment or views of the paper on specific issues within its agreed editorial policy; and
- the right to appoint or dismiss journalists and to decide the terms of their contracts of employment within the established policy of the organisation, and the right to assign journalists to stories (McGregor, 1977: 155).

The Royal Commission acknowledged the difficulty of defining the limits of the proper responsibility of editor and proprietor and that 'editorial and managerial decisions were inseparable' (ibid) and that, therefore, the editorial policy could not and cannot be separated from the financial success of a media enterprise. Nonetheless, an explicit codification of editorial and journalistic independence along the lines of the catalogue of 'rights' defined by the Royal Commission is necessary if the goal of a pluralistic, independent, iconoclastic and investigative media is to be achieved. In the first instance we believe that consent for further mergers and concentration of ownership of the media in the UK should be conditional on a strengthening and extension of editorial and journalistic independence. This may seem unnecessary in the light of the measures proposed by the Government. Accordingly it is necessary to consider how the potential harm that might arise from such a concentration of ownership could be mitigated. That is why media policy and regulation must consider conduct as well as structure and why attention should be paid to these neglected recommendations of the last UK Royal Commission on the press.

If the government wants to put in place an ownership framework that permits consolidation and integration, it must promote editorial freedom and break the link between ownership of media corporations and editorial control. Looking to other countries, we can see a range of safeguards and guarantees on media pluralism. Professor Peter Humphries recommends the following:

> There could be one law to guarantee editorial independence across the board. Legislation might provide for editorial codes of practice or company statutes that would protect journalists' editorial integrity and independence and guarantee them a degree of industrial democracy. A new law could introduce a 'clause of conscience' which would increase protection of journalists' editorial integrity by making it costly for owners to sack them, as in France. Austrian journalists have the right not to collaborate on a piece of work that does not reflect their private opinion. In Sweden, journalists can refuse to write a story that offends against a code of conduct agreed between the publishers organisations and the journalists' union (Humphreys, 1996 pp108-109; Humphreys, 1994 pp104-111).

In the field of broadcasting, a degree of internal pluralism and editorial independence from proprietorial control could be guaranteed by providing for pluralistically-composed independent programme advisory councils within the largest, most influential commercial broadcasting organisations (another idea from Germany).

In sum, whether by statute or by codes of practice, ways must be found to limit proprietorial interference, to separate financial from editorial decisions, and generally to promote high professional standards of journalism and the goals previously realised through ownership controls. (Gibbons, 1992).[3]

Conclusions and IPPR proposals

The government has the opportunity to strike a deal: OK to some ownership consolidation, but only if you accept legal guarantees of media democratisation, pluralism and diversity.

Progressive democrats have tended in the past to resist a debate about media pluralism. If they continue to do so, they will lose the opportunity to shape a plural broadcast media in the transition to digital. The danger of that lack of engagement are either that ownership rules will remain unreformed and domestic players will be unable to compete with global media giants, or that ownership rules will simply be removed with the naïve hope that general competition rules will somehow miraculously deliver pluralism. The question is not whether competition will deliver pluralism, but how to put in place the competition framework, and other measures, with will deliver pluralism.

If large media companies fail to engage in a constructive debate about media democratisation, there is no way beyond the impasse. Those who seek to relax ownership rules claim that there is no link between ownership and editorial control. If that really is the case, they should be prepared to permit their journalists the rights that will formalise that approach. This is not about cumbersome meddling in the internal affairs of broadcasting organisations, but about increasing the autonomy and responsibility of the journalistic profession.

The basic challenge for competition regulation is clear: digital presents the possibility that small content creators can access audiences even if they do not own the infrastructure necessary to reach them. To make this possible should be the ultimate aim, and open access rules should be the instrument. If regulation is weak and flatfooted, however, it will fail to promote that access and the big vertically integrated players will control content, networks and audience. The approach to pluralism should be 'safety first'. Build the necessary competition regime for the long term with pluralism in mind, maintain existing rules for mass analogue services, and promote more imaginative measures to guarantee journalistic independence.

The objective of media pluralism and diversity has been crucial for the analogue environment. With the shift to digital delivery, the objective remains crucial but the instruments for its realisation will change. IPPR recommendations fall into four categories.

Ownership regulation

- Continue and support existing ownership rules for mass audience analogue broadcast services.
- No shift in methodology of ownership and pluralism measurements.
- No relaxation of existing market share limitations or cross media ownership rules without media democratisation reforms
- Retain some support for independent production quotas

New competition rules for the digital environment

- Support for the measures outlined in the report to the Council of Europe, 'Pluralism in the Multi-Channel Market'. (Marsden 1999).
- Regulation for competition/ mergers regime should operate with a clear view of the goal of promoting pluralism through public interest tests.
- Monitoring and measurement of concentration in new media markets should be continued.
- Strong regulation of bottlenecks, gateways and open access, CAS, EPG.
- Whilst economic thinking suggests that short-term monopolies might be tolerated in the information sector if they are not abused in terms of competition, it must be made clear that abuse of that position includes abuses that result in a decline in pluralism.
- In public interest tests for mergers in the sector, impact on inhibitions of source, content or exposure pluralism should be taken into account.

Balancing internal and external pluralism through continued support for public service broadcasting

- Continued support for positive content regulation: public service broadcasting to provide internally diverse and plural programming.
- Continue C4 in public ownership to ensure a second internally diverse public service channel.
- Must carry/must supply rules applied to Public Service Broadcasters.

Media democratisation reforms: Breaking the link between ownership and editorial control

- Media companies arguing for relaxation of ownership rules should be invited to suggest new measures to promote editorial independence; and provide internal democracy and transparency.

- Regulators should examine how such commitments should be enforced, and how the guidelines of the Royal Commission on the Press could be extended to broadcast services.
- New clauses in employment contracts of journalists to guarantee editorial independence.
- Larger broadcasters (defined by share of the market) should be called upon to develop codes of practice for pluralism and pluralistically composed broadcasting councils.

Endnotes

1 Committee of Experts on Media Concentrations and Pluralism (MM-CM) (1997) *Report on Media concentrations and Pluralism in Europe* Strasbourg: Council of Europe 20 January.

2 For an astute analysis of how the web is developing its own forms of media agenda power, see Richard Rogers, *Preferred Placement* van Eyck J, Akademie Editions 2000.

3 I am grateful to Peter Humphreys for research in this section.

3. Content regulation: a new settlement

Liz Forgan and Damian Tambini

The background

Since the emergence of broadcasting in the last century, there have been elaborate rules and procedures, over and above the criminal law, to regulate media content. These have been both negative – to restrict distribution of certain information, text, sound and images to screens and amplifiers, and positive – to promote access to content.

Regulation has been organised according to the technology used to deliver content. There have been different rules for the regulation of television, radio and film and according to use patterns and consumer expectations.

Content regulation has generally been applied to those channels of communication that are accessed by children, and those that are deemed public, influential, pervasive and invasive, including premium rate telephony, broadcast radio and television.

The instruments of regulation include direct statutory requirements, compulsory powers given to a Regulator, franchise or licensing requirements and statutory or discretionary Codes of Practice.

Historically, some regulation – notably of broadcast television – has taken place by systematic scrutiny and in advance of publication. Increasingly, liberal argument and the sheer volume of output have brought about a shift toward post-transmission, complaint-based regulation.

In addition to statutory forms of regulation there are instances of self-regulation voluntarily undertaken by providers as a source of competitive advantage. These are outside the scope of this paper but are significant as part of the total regulatory effect experienced by the consumer of communications services.

Why legislate now?

Although the precise speed and pattern of the development of digital media markets cannot be forecast, their very arrival constitutes a watershed. The 1996 Broadcasting Act foreshadowed the need for a radical approach but reflected widespread uncertainty about the way revolutionary new forces would take the industry and left important options open. This Bill must grasp the nettles.

Technological convergence makes it difficult to regulate according to delivery technology because the definitions of services become blurred and regulators/regulations based on technology come into conflict and overlap. It becomes possible to separate the regulation of *carriage* –the networks, transmitters and infrastructure of delivery – from regulation of the *content* that is delivered. Whilst discussions of content cannot be isolated from other areas of media policy such as competition and ownership, the aim here is to focus on content: to clarify the objectives, rationale and instruments of content regulation and how they may change in the coming years.

Some aspects of the need to regulate communications – and the capacity to do so – have historically derived from spectrum scarcity, notably in respect of broadcasting services. Digital technology and signal compression have transformed spectrum availability and calls to relax regulation have followed.

Regulation of the content of communications can be both an inhibition and guarantee of freedom of information and expression. As such there is a case for continued regulation even in a mature and plural market place but it requires specific and contemporary justification.

As spectrum scarcity for radio and television is replaced with plenty, and the number and flexibility of services increases, the pervasiveness, invasiveness, publicness and influence of channels decline. So, therefore, do some of the justifications for regulation.

At the same time the role of media and communications in society is shifting. Television is no longer merely a source of entertainment and information. With the switch to digital it will increasingly become an important gateway to public services and a key element of the nation's infrastructure. In such a context there is a need for a new definition of the public interest in communications which takes into account all these aspects of the role of communications media, without neglecting the validity of any of them.

The marketplace is in a highly volatile condition. The legislation must reflect the central fact that we are regulating for a process of transition in which the regulatory environment should act as enabler and facilitator of change, taking into account the significant novel public benefits offered by new services. New digital services offer important advantages such as interactivity, access to a huge range of free cultural resources, any to any communication and cheap publishing. Regulation should incorporate these new policy objectives into its overall approach and not merely protect older broadcasting principles. We cannot be sure how the market will develop because consumers have never before been confronted with these kinds of choices and this range of digital services including the Internet.[1] Regulation must, however, reflect the new choices, because they will impact upon key issues such as child protection, pluralism, and so forth.

What assumptions about 2015?

Whilst it is not possible to predict exactly where the market is going, it is possible to make some assessments of the parameters of change.

- Analogue television signals will be switched off.
- Spectrum availability – as opposed to cost or the criminal law – will not be a barrier to market entry in audiovisual services.
- Spectrum will not be free to anyone. It will be traded for cash (UMTS) or public goods (public service broadcasting, other social purposes) or some mixture of the two.
- Although spectrum licensing will decline as a means of regulating, some core broadcasters will still benefit from must carry and prominence in EPGs and can therefore be expected to undertake public interest obligations in return.
- Access to the web – in some form and via some form of media – will be near universal.
- Content regulation will continue to be a matter for national governments while competition and standards will increasingly be dealt with at European or global level.

- Consumer control over content is increased with use of CAS, filtering software and TiVo.

Can we regulate?

The driver of radical change is the World Wide Web. The government has proposed that every citizen should have access to the Internet by 2005. If this aim is realised, the vestigial regulatory regime that applies to the Net must – at least in terms of negative regulation – inevitably become the initial benchmark for regulation of other delivery platforms.

As the Internet's capacity to deliver video, audio, text and pictures improves, audiovisual services delivered over the Internet may displace audience share of other platforms. There is already evidence that Internet use is eroding television audiences in the US. If Internet services are delivered to digital televisions it may eventually be difficult for consumers to distinguish between television and Internet services, and radically different regulatory environments may be difficult to defend.

That does not mean that all media must share the same level of regulation. It does throw a salutary bucket of cold water over any hope that regulation of the content of communications media can be just a modified version of what has gone before. The constant reminder of the global anarchy of the Net is an excellent check to bureaucratic nannying.

It requires us to show cause wherever it is proposed on any other medium to restrict material which would not be restricted on the Net. And material that *cannot* be regulated on the Net, however desirable it might be to do so, should only be regulated elsewhere if regulation can be effective.

However, although we are at a transforming moment, we do not start from a blank sheet of paper. British consumers are used to intervention in the content of broadcasting, film, video, press and certain areas of telephony. There is no public appetite for a sudden lurch to a free-for-all.

The challenge posed by unregulated cyberspace provides a spur to sorting the genuinely valued and necessary features of regulation from the accretions that have grown up in an ad-hoc way since the beginning of mass communication. That process must involve new forms of democratic participation.

The costs and pitfalls of regulation

Currently, eight bodies have responsibility for content regulation. Leaving aside the BBC board of governors, the Home Office and the DCMS, the total direct cost of running the BBFC, BSC, ICSTIS, ITC, the PCC and the Radio Authority in 1998 as measured by their expenditure was in excess of £29m.[2] Some sections of the industry are subject to two or more regulators or to a co-regulatory regime. As the process of convergence continues, these regulators are likely to come into increasing competition and conflict.

Multimedia content is increasingly seen as an industrial product as well as a cultural artefact. The global marketplace for content and particularly television programming is expanding and content policy must take that into account. According to a recent government-sponsored report, the UK's television industry UK is running an expensive trade deficit in television. The total value of UK exports of television and radio in 1996 was £234m; whereas the UK imported television and radio programming worth £516m (see also Graham 1999; DCMS 1998). Though the figures have improved in more recent years, there is a consensus that UK content industries could perform much better. A significant body of opinion holds that content regulation, which imposes a non-economic notion of quality, is partly responsible for that under-performance. It remains to be proven however, that content regulation is responsible for low sales of UK content in the global market (DCMS 1999). Clearly there is a need to balance these economic concerns with those of the public interest in content of expected standards, and these concerns should be taken into account in discussions of any new regulation.

Content regulation: a reassessment

In order to chart the territory of content regulation over the next decade, it is necessary to return to the basic rationale: this applies to both negative and positive content regulation. It also applies to those forms of regulation that do not easily fit into either category, such as those regulations that seek to ensure a plural, impartial, democratic media.

In the broadcasting environment, as elsewhere, markets are excellent mechanisms for delivering services tailored to consumer demand on the basis of individual decisions in competitive markets. They do not always lead to the optimum outcome in terms of public provision or social benefit. The regulation of broadcasting markets has long accepted that there are public and merit goods that will not be delivered by the sum of individual decisions in a perfect marketplace, and that regulation, subsidy and other forms of intervention may be justified.

British users of broadcast services, film, video, premium rate telephony and, to a lesser extent of newspapers, are accustomed to regulation both negative and – in the case of broadcasting – positive, and regard it as a guarantee of choice, quality, fair dealing and safety. A sudden and comprehensive relaxation would command little or no support.

Consumer expectation varies according to the degree of pervasiveness and invasiveness of media. Paid-for entertainment services, whether subscription, pay per view or others, have, and are expected to have, lighter regulatory conditions than those available free-to-air. Broadcasters with smaller shares are less regulated than channels with mass audiences. Live premium rate phone calls are regulated to a different standard than other premium rate services. Top shelf publications are expected to contain content that would not pass the codes applying to mainstream publication.

Convergence will alter the future meaning of categories such as private/public communication and top shelf/mainstream publication. Future regulation must respond to those shifts and consumer attitudes to the communications transformations they bring.

Negative content regulation

The huge edifice of negative regulatory control over taste, decency, offense, and consumer protection is the area in which most radical change must inevitably come.

General principles:

There is a clear distinction between protection of adults and of children. In democratic societies we regard adults as subjects of basic freedoms of speech and movement. Children's freedoms we routinely restrict for their own safety. Parents and legal guardians should be empowered to protect their children and informed of their responsibilities.

- The freedom of the Net should not be compromised. Negative regulation should encourage access to interactive media hardware and the broadest possible content for readers, watchers, speakers, publishers and listeners.
- Users will learn over time how best to use the new media by developing informal rules and a culture of responsibility. Public authorities have an important role in encouraging that process.
- Illegal content should be dealt with by a fast and efficient public reporting system.
- Content labelling, rating, filtering, censorship and selection should be done in ways that are transparent and accountable, according to publicly agreed guidelines to avoid abuses of filtering and unjustified restrictions on communicative freedom.
- In a fast-changing environment, the definitions of what is illegal or inappropriate to deliver electronically will change rapidly according to use patterns. Regulation must be flexible and responsive.

Harm to children

The desire to prevent harm to children has been a central driver of regulation, mainly on broadcast television and in film and video. It has traditionally operated to prevent children from being exposed to violent or sexually explicit material unsuitable to their ages, on the assumption that children habitually used television and that their parents could not always be expected to supervise their use. In the long term the public, with the aid of filter software, should be able to regulate its own content and chose service packages that deliver the content selection that they prefer, for example for children. In the short to medium term, however, there remains a strong public expectation that existing services will continue to be regulated. Regulation may be imposed on Internet Service Providers but the nature of the Net means that this can only be complaints-based in that it cannot be comprehensive or even consistent, and is likely to have detrimental effects.

There is, however, clearly still a public desire for and interest in regulating to protect children on other media despite our powerlessness in respect of the Net. The Watershed on terrestrial television has proved a simple and practical compromise between the freedoms of an adult audience and the desire of parents to retain some control over their children's viewing.

Harm to adults

A combination of the criminal law, trading standards and consumer education must eventually replace detailed prior examination of content by regulators. Prior restraint is not a sustainable principle. *Caveat emptor* and an active consumer movement protect buyers in the high street and must, when the novelty has worn off, also protect buyers on the Net or over the phone.

It is open to any media organisation or group to elect to follow more restrictive standards as a matter of commercial advantage but statutory intervention should be phased out as market share and consumer attitudes change.

Harm to society

It is arguable that society at large has an interest in not permitting the sensibilities of its citizens – especially 'vulnerable' ones – to be coarsened by constant exposure to violent or sexually explicit material. The same could be said of controls on technical information eg on bomb making.

Here the existence of the Net makes regulation elsewhere unsustainable. Those likely to be damaged by ugly material or to be seeking criminal expertise will be most certain to find it on the Net. Regulation on this count would be unjustifiable on the grounds that it could not achieve its object.

Fair trading

The ITC, which currently regulates broadcast advertising through its advertising code, has indicated that the sheer volume of advertising in a digital interactive environment and the blurring of boundaries between linear services and levels of interactivity will create problems for the current model. Indeed there is a general need for some rethinking of fair trading regulation given convergence. As for other areas of complaints-driven regulation, it is crucial that the public are made aware of their responsibilities as complainants, and of where and when regulation is being altered or removed. In the long term a fair trading authority or general advertising regulator rather than a communications regulator should be responsible for these aspects of negative regulation.

Public and private: when should regulation apply?

As distinctions between existing technologies and between public and private media become less clear, there is a need for a new guidelines for deciding when rules apply according to use and consumer expectation rather than by delivery technology. The arguments for monitoring and rating content are strongest where:

- Texts and channels have mass public reach,
- Material is specifically directed at children(Tambini 2000).

In the short term regulation should be applied to core services in broadcasting. In the long-term this will require ongoing research on the use patterns of new media, and public expectations and attitudes.

A health warning: fostering a responsible culture for new media

'You are now leaving the regulated sector. Your safety is your own concern'.

Public awareness of the dangers of new media, and responsibilities for protecting children are far too low.[3] Parents should be made aware of their responsibilities, perhaps via information included in software, for their children's welfare in relation to new media. Parents should not be left to their own devices in this respect since the technology may be too difficult for them to master. A public information campaign should coincide with the launch of a rationalised complaints and monitoring strategy, and should include:

- Information at point of sale. Of Internet software, and also of digital TV, mobile devices and other web-capable devices.
- Media education in schools including items on when to be sceptical of the trustworthiness of online content.[4]

- A 'health warning' for users accessing unregulated content through other, more regulated platforms.
- Deterrents should encourage responsible use of the Net. There should be a new offence of deliberately directing harmful content to children, for example, by malicious meta-tagging.
- Public awareness of the consequences of distributing harmful and illegal content is as important as a deterrent. But this should not be conducted in ways which dampen the Net, by making ISPs or content providers over-cautious and restrictive.

Positive content regulation

General principles

In a multicultural, multinational, regionally diverse society it is increasingly difficult to specify a central cultural canon or authority which broadcasters have a responsibility to deliver to all.

- As genuine consumer choice expands, an elitist view of culture becomes more difficult to defend. Positive content regulation should not promote one culture or ethnicity.
- However, as government services are delivered increasingly by electronic means, there remains a clear public interest in universal provision of certain civic content such as whether this is provided by a competitive market or by regulation.
- As information and communication technology becomes more important to more aspects of citizens lives, then the range of services to which citizens should have guaranteed access broadens rather than contracts to include government services, (such as education) and government information, a government portal.
- The range of cultural, entertainment and information services available to UK citizens in the future should be no narrower than those that were available in the past.

Many argue either that these can no longer be guaranteed by regulation in the new environment or that they are no longer justifiable, as technology offers consumers greater choice and control. Clearly in a pull model of media provision rather than the push model of scarce broadcasting, the tools change, but there is a role and potential for regulation to guarantee reasonable and affordable access to those services that should be available as of right to citizens.

It is crucial that general and consistent principles are developed that specify which content should be made easily accessible free to all. In the past, must-carry rules, event listing in broadcasting and universal service in telecommunications approximated a consensus on communicative entitlements. In the new communications environment, not only access to public broadcasting, but access to electronic public services, to government information, and perhaps even e-democracy capabilities such as polling and voting might be included in the basic package of services to which citizens should have access. The rationale for this comes not only from the need to promote social inclusion, it also relates to needs to deliver government services efficiently and the need to promote rollout of the most efficient infrastructure for e-commerce.

Democracy, pluralism and the media

Votes should not be bought and sold. We do not want our political information to come only, or in large part, from a handful of commercial providers. Neither do we want to be fed mainly or wholly by information provided by government. In either case we are worried about a powerful minority putting their interests before the interests of the public as a whole, and undermining the ideals of freedom, equity, fairness and self-determination upon which our democratic institutions are based. There are therefore rules to ensure that news and information – especially news and information about government and politics – is accurate, of high quality, and presented in a way that is impartial. If it is not appropriate to regulate for impartiality, care is taken to ensure that a variety of different sources of information can be accessed.

According to the Council of Europe Committee of Experts, the benefits expected from media pluralism are:

to serve democracy by providing citizens with a broad range of information and views needed for the effective exercise of citizenship; to give minorities the opportunity to maintain their separate existence in a larger society; to limit potential social conflict by increasing the chance of understanding between potentially opposed groups and interests; to add to the richness and variety of cultural and social life; to open the way for cultural and social change (Council of Europe 1997).

ITC guidelines, including quality news provision, ITC due-impartiality rules, BBC due-impartiality rules, rules governing accuracy, plurality and diversity, documentary and current affairs quotas, party political broadcast obligations and privacy codes are examples of content regulations that ensure that the powerful broadcast media support, rather than undermine, democratic communications.

Competition and pluralism

There is a strong argument for the removal of ownership restrictions and restriction on advertising and viewing share, to permit consolidation and vertical integration in broadcasting and communication industries. There is also a strong argument that a competitive market in a situation of channel and programming plenty will eventually deliver such choice to the consumer that specific regulation to promote plurality will no longer be necessary. However, we are not there yet. Competition rules alone would not deliver sufficient pluralism of voice. If consolidation is allowed to occur, it should be balanced by more stringent obligations to ensure a plural and democratic media – for example on internal pluralism, impartiality and specific measures to promote editorial independence. Such measures must be put in place to ensure that the broader public interest is not eclipsed either during the transition to digital or in any end state.

Ownership rules, in addition to acting as a structural guarantee of competition, have generally been the key means of promoting media pluralism. As competition policy develops in the digital environment it will move away from structural competition rules – which were set at high thresholds – and focus on behavioural regulation, including the regulation of bottlenecks and gateways such as CAS and EPGs to ensure that consumers can access a broad choice of content.

It is, however, difficult to agree on an objective measure of when a media market is sufficiently plural or diverse. Policy develops rather by outlining what we wish to avoid: a media landscape in which an individual or company, foreign or domestic, can control a significant proportion of the information entertainment and opinion that make up citizens' media diet.

A consistent and transparent regulatory framework

Clearly, content regulation will be transformed in the coming decades. In the future, consumer choice coupled with competition may deliver much of what was in the past delivered by content regulation. However managing the transition to that situation remains a tricky process, and the costs of a mistaken strategy could be very high. In the shorter term it is necessary to come up with a workable transitional solution. Consumers expect content regulation and content regulation has structured the market for some time: a sudden relaxation of rules would result in the undermining of existing public institutions and widespread public dissatisfaction. In order to formalise inconsistent ad-hoc regulation, a new contract with broadcast licensees must be reached.

Clearly central control of market entry is in long term decline, but this should be a managed, gradual process rather than a sudden one, and should be conditional on shifts in consumer expectation and the market. Over the next ten years the principle of detailed regulation of all broadcasting should be replaced with one that formalises the existing de facto sliding scale of regulation according to the degree of consumer expectation of regulation and market share. At one end of the scale lie the public channels that operate in a contract with the public, delivering certain obligations in return for the privileges of prominence, spectrum and the kite-mark offered by government regulation. At the other end lie unregulated services delivered to niche audiences.

There is clearly a necessity for a new settlement. At present there is a severely uneven playing field, which is governed by no general objective principles. ITV has substantial regulatory impositions while Sky's are minimal. Channel 4 has free spectrum. Channel 5's is paid for in cash and it has to reach a minimum quality threshold (see also ITC, http://www.itc.org.uk). Commercial radio pays a combination of cash and content obligations. Telephony carries no positive regulation and its spectrum is valued differently from commercial radio. UMTS was valued by means of auction. Other prices are fixed by estimation. The allocation of spectrum between media is not transparent. These examples illustrate the need to introduce objective criteria for the content element of the trade for spectrum as well as for the financial criteria.

There is therefore a continuing need for content regulation, but a pressing need to reform the current system that is being overtaken by market and technological developments. Citizens/users expect that particularly the core channels will continue to be regulated, and there is considerable evidence that the market alone will not deliver what is in the public interest. Although they are losing control of market entry, government and regulators do still have some cards to play in the transition to digital. They control some premium content in the form of government services, and can through licensing still confer prominence through EPG regulation and must-carry rules.

What are the options for a regulatory structure to govern content during this transitional time?

The requirement is for a regulatory structure capable of:

- Simplifying the complex historical tangle of regulators and regulatory mechanisms;
- Delivering content objectives consistent with consumer expectations;
- Achieving fairness, consistency and minimising the burden for competitive industry;
- Moving with any pattern of convergence, growth or substitution thrown up by this volatile and unpredictable new market place; and
- Facilitating the development of an open, inclusive information infrastructure accessible to all.

In broad terms there are six possible approaches:

- Radical deregulation of content: trust to the free market subject only to competition and fair trading regulation.

  Risk: short term chaos, possibly permanent inability to deliver objectives, consumer outrage, downfall of the government of the day long before any effect of a mature market place could be observed.

- Seek a pragmatic means of continuing something like the current pattern of regulation but with an unspecified 'lighter touch'.

  Risk: unsustainable without some principled underlying clarity as to the meaning and progress of 'lighter touch'.

- Confine regulatory intervention to a few designated public service providers, eg the BBC and Channel 4, protected by spectrum and funding privileges. Lift all special content regulation from every other service.

  Risk: a less violent version of the first approach, as the whole of British commercial radio and television lurched immediately into a free for all.

- Make no advance judgement as to the likely development of the communications market. Reduce content regulation to a clearly stated set of minimal principles and grant substantial discretion to a regulator to interpret them freely as the market place develops.

  Risk: impossible uncertainty for providers and consumers alike and would be unacceptable by modern standards of accountability.

- Embrace one or more scenarios for the development of communications through the coming decade and regulate to achieve or to accommodate to those scenarios.

  Risk: being locked in to a set of wrong assumptions producing an inappropriate regulatory regime.

- Develop a trigger-based regulatory escalator designed be consumer-focused, applicable to any medium, flexible enough to move with any market movement and sufficiently objective to be transparent.

  Risk: complexity and difficulty of agreeing operational measures. This is a new approach we propose for detailed consideration.

A proposal: the CEMS index

Subjective judgement can never be wholly eliminated from a task as complex as the regulation of the communications industry. However, at a time of intense global competition it is important that such regulation is as transparent, certain and objective as possible if providers are to play on level fields and consumers are to have confidence.

It is no longer possible to regulate according to technology, and to clearly distinguish invasive mass media where content regulation applies from private media where freedom of speech concerns prevail. Regulation for the transition must incorporate some form of sliding scale along the continuum between the two, and be based on use and consumer expectation rather than technology.

How then to devise consistent and fair principles to decide when detailed content regulation should be applied? We propose a single graded classification of communications services irrespective of delivery method that would take account both of historic and of developing circumstances. The key measures of invasiveness, pervasiveness, publicness and influence are Consumer Expectations and Market Share (CEMS). Content regulation should be applied in part on the basis of which remaining privileges the content provider enjoys (spectrum or otherwise) and in part according to what we call the CEMS index.

Both consumer expectation and market share will change in unpredictable ways in future as convergence, substitution and other market factors – such as analogue switch off – play out.

At any given time, therefore, the index would provide a rating for any service consisting of the product of the two indicators. This would determine into which of the broad regulatory bands the service would fall. All services could move between bands with 'promotion' or 'relegation' decided by periodic re-ratings to be carried out at the discretion of the Content Regulator. Services must be categorised as a single entity and not by piecemeal examination of particular programmes or strands of service.

Different regulatory regimes will apply to each category as shown in Table 3.1.

Table 3.1 CEMS Scale: indicative content

Division	*Examples of content rules that could be applied*	*Qualification criteria*
Group 5 eg Net	None	*Privileges:* none *Market share:* insignificant *Consumer expectation:* low
Group 4 eg New Linear services	Taste and decency – by classification alone	*Privileges:* none *Market share:* above specified minimum eg 5%. Capable of UK regulation *Consumer expectation:* low
Group 3 eg ITV, C5, INR, ILR	All the above plus: • Specification of a range of programming • Due impartiality • Civic information • UK origination	*Privileges:* prominent must-carry or EPG position. *Market share:* significant *Consumer expectation:* Moderate
Group 2 eg C4	All the above plus: • Education • A range of programming broader than the market provides and having regard to specialist as well as mainstream interests. • Impartial news and information programming including core civic services • Freedom to take risks – with new talent, new forms and new services • A commitment to UK-based talent and content. • Innovation • Information in peak • Taste and decency • Protection of children via a watershed	*Privileges:* commercial protection, secure spectrum, EPG prominence, must-carry. *Market share:* > 8% of total audience. *Consumer expectation:* high
Group 1 eg BBC	All the above plus: • Taste and decency • Representation of regional/minority cultures • Systematic provision of education services • Entertainment and information services of high quality free to air to compensate for the divide between techno-rich and poor • Requirement to sustain a minimum reach as a guarantee of future funding. • Easy access to electronic public services. • Non-commercial content search engines.	*Privileges:* licence-fee supported services. privileged/secure spectrum; must carry, prominence on EPGs. *Market share:* more than 15% of total audience. *Consumer expectation:* high

The market share indicator is the easier to measure though there is a decision to be made as to the definition of the relevant market for the purposes of calculating share. These should be consistent with measures of ownership for competition purposes.

But if we are to retain the values and virtues of the British communications landscape to date- and take proper account of changing attitudes to, say, telephony as its communications role changes, market share alone – the measurement of the accumulated choices of individuals at a particular moment – is not enough. It is imperative that it is checked against other expressions of the wider attitudes and expectations which people have of an industry which has a special role in personal and national life.

And if those wider consumer expectations are to have equal status with the clarity and power of market indicators they must be systematically examined and authoritatively established. Traditional research methods such as currently used by BARB, RAJAR and special surveys, together with complaints logs would constitute the principal data. There could be a role for a Consumer Communications Council or citizens' juries to act as the focus for and articulator of the views of all communications users.

In the long term this more flexible system could resemble something like a rating on demand or kite-marking system.

The BBC

Currently the single most powerful feature of positive regulation in the UK communications marketplace is the funding by license fee of the BBC, followed to a lesser extent by the commercial protection of Channel 4. The *quid pro quo* for both is an array of public service obligations which must continue to be redefined as other circumstances change and consumer needs develop (see Table 3.1)

Decisions about future funding and public service obligations for the BBC will be made having regard to the movement of the index.

Two changes to the Charter are, however, needed with reference to the content performance of the Corporation:

- There must be separation of the commercial businesses from the public service operations with completely transparent trading between them taking account of quantified value of the brand and of licence fee investment.

- The Governors' responsibility should be confined to guaranteeing independence from government and accounting to the licence payers for the public service performance of the Corporation only. All other regulatory powers should be transferred to the communications content regulator (see below).

Who regulates?

The question of how many regulators and who they should be cannot be settled with reference to content considerations alone.

However current developments for content regulation do have direct implications for regulatory reform. First, we are moving to a situation in which public complaints, and notice to take down play a larger role. Therefore, the process of complaining should be rendered more transparent and simple to the consumer, and this would suggest that there should be one single address for complaints and a more efficient way of recording and dealing with them.

Second, convergence is leading to conflict between regulators. There is a strong argument for merging existing content regulators into one body in order to avoid counterproductive turf wars and conflicts between regulators during convergence and in order to make more transparent, efficient and accountable a complex and opaque system.

Third, the proposal for a single regulatory index would require a single regulatory regime for all content services – as would the tendency of the market to convergence. As the difference between online games for example and linear content becomes less clear, it will be convenient to focus technical regulatory expertise in the field into one body.

Content regulation must operate in close relationship to carriage and competition regulation but will often drive in different directions. If balances are to be struck with due regard to the consumer interest in content regulation there are powerful arguments for a separate content regulator. Conflicts would be resolved by negotiation between the two regulators, each having regard to its primary responsibility. In the event of irreconcilable difference between competition regulator and content regulator, final resolution should be a matter for political settlement at Ministerial level.

We therefore argue strongly for a single content regulator. The eventual fit between content and economic regulation needs more research. From the point of view of content regulation, what is crucial is that the public interest in content regulation is given the priority it deserves even as industrial policy concerns become more important. Should content regulation fail, there will be a political price to pay. Ensuring that content regulation is based on a fair and transparent contract with providers, and that genuine and powerful economic sanctions can be brought to bear to enforce content regulation is essential.

Endnotes

1 We should not however overestimate the speed of change. Despite the availability of multi-channel television for over a decade only about 35 per cent of homes have yet adopted it and even in these homes, 60 per cent of viewing remains with the terrestrial channels. (ITC 2000). This study does not however cover Internet use.

2 Sources: Tambini (2000b); *Regulating Communications* Green Paper 1998. This represents current expenditures of regulators, however, and does not include the costs to the economy incurred as a result of market distortions. Having the obligation to provide certain content at subsidised prices can be, in purely economic terms, a wasteful use of resources and impose a cost on the economy. The extent of such a cost is, however, extremely difficult to measure. Safety and quality for viewers stimulates usage of media services by customers who would otherwise have been deterred from consumption. Accordingly, content regulation can improve the market conditions.

3 The Internet Watch Foundation report that 18 per cent of Internet users in 1999 claimed to have heard of the organisation, which is responsible for Internet regulation, and relies entirely on user reporting and complaints.

4 Existing initiatives include the BBC's Webwise, and Childnet International. See www.netaware.org.

5. Modernising communications regulation

Damian Tambini and Clare Hall

From threat to opportunity

UK regulators have led Europe and the world in terms of telecomunications competition and liberalisation, and in terms of the development of public broadcasting standards that protect minors, promote pluralism, and promote wider access to communications services.

It may be asked why a progressive government would seek to reform such successful institutions. The answer is simple. Although the existing communications regulators have done an excellent job, the technological and market changes associated with digitisation and convergence are undermining existing institutions of regulation.

There is now a widespread perception that the shift to digital media is a threat to the established structures and principles of public broadcasting and telecommunications regulation. The role of regulatory authorities defined in relation to technologies of broadcasting and telephony for example becomes less clear when technological convergence blurs the boundaries between those technologies. Consumers can now receive a range of communications services via a variety of different networks and appliances with the effect that previously distinct markets are merging (for example, voice telephony delivered by wire, cable, or radio). In this context, regulators defined in relation to these separate technologies are finding increasingly difficult to protect the public in terms of the traditional goals of encouraging fair trading, consumer protection, and a quality, safe democratic media. And both consumers and the industry face confusion about who, ultimately, is responsible. But this is not only about threats to the public interest: new technologies also present entirely new positive potentials which should be exploited if new media are to be a force for a democratic and inclusive society. Regulation may be necessary to ensure those potentials are realised.

The current moment should be seen in a broader historical perspective. In the 1920s, the UK responded to the new technology of broadcasting in a way that contrasted sharply with the approach of the United States. Government-controlled spectrum was given over to the BBC, initiating a system of publicly accountable, publicly controlled broadcasting that is still in place. Telecommunications have a more complex history of alternating state and commercial ownership over 150 years, but the sector has recently followed a trajectory similar in some ways to broadcasting: a state monopoly run by a semi independent centrally funded body giving away to a commercially funded organisation and then to ever increasing competition and innovation in a liberalised market. The challenge to government must be to work with industry and consumers to set out the conditions for a new media ecology for the digital age that is socially inclusive and economically dynamic. Regulatory reform should be seen as a means to guide the process of change. Should this opportunity be lost, the consequences will be felt well into the future and to the detriment of consumers, industry, society and citizens. The core problem, as the head of telecoms at the OECD put it, is 'how to facilitate the process of convergence or at least to ensure that regulation does not present barriers to convergence' (Ypsilanti and Xavier/OECD 2000 p644). At the same time, broadcasting institutions must be both fostered and prepared for the new environment.

There is already a consensus that the current system of law and regulation needs reform. From one angle, the burgeoning industrial sector is unable to take business decisions in the absence of clear leadership by government and regulators. From another, consumer and citizen advocacy groups are concerned that competition and commercial pressure will not deliver the full range of social benefits that are the potential gain from the digital revolution. All the stakeholders recognise that the current regime is increasingly ineffective.

Whilst regulatory structures in the analogue era have been effective in helping the media and communications sector perform well in the United Kingdom and worldwide, the cracks are beginning to show. The obsolescence of old sectoral barriers questions the number and the role of the regulatory bodies and the degree of inconsistency and overlap in both regulatory objectives and regulatory process. The old regulators are often deemed too bureaucratic, too slow and unresponsive to the new environment, inhibiting commercial and technical development within the industry. Whilst we should not expect communications markets to be transformed overnight, and many aspects of regulation will remain the same, new digital media cannot be regulated in the same ways as were analogue broadcasting and telecommunications.

This pamphlet reviews the rationale for reform of communications regulators and then considers some of the options open to Government. The challenge to policymakers is to lay the groundwork for progressive reforms that remove unnecessary regulation, modernise the current system, and promote the broader public interest in the new communications environment. It would be difficult to overestimate the importance of getting this right. Our communications system is the space where our society thinks, debates and expresses itself: with the move to more information based economy it is also a key aspect of our industrial infrastructure.

Progressive principles for communications regulation

Radical reform of communications regulators has been on the agenda for over a decade and has been the subject of copious research and debate. According to William Melody,

> Regulators must be able to adapt to changing circumstances or they will simply become a bureaucratic drag on industry development. They must have the power to interpret policy and adapt their approach and methods accordingly. In a dynamic industry sector, what are needed are fewer, stronger more independent regulators with responsibilities for a proactive and forward-looking approach to regulation (Melody, 1997b: 24).

Back in 1997, the director of the ITC noted that

> four or five separate regulators with increasingly overlapping responsibilities is an accident of history which cannot make sense for the future... The work of the BSC overlaps with and duplicates that of the BBC, S4C, and the ITC. For example, the content of the BSC's code covers the same ground as a

> part of the ITC's Code and the equivalent BBC and S4C documents. Broadcasters must therefore satisfy the different terminology of two separate documents produced by two separate bodies in relation to a single programme issue. This cannot be of service to broadcasters or the public in terms either of clarity or resource use, particularly as the BSC is funded out of public expenditure. (Rogers 1997)

We should take on board the basic line of reasoning used by the then director of the ITC, and carry it forward into a converged world. Significant reform of communications regulation is now official policy of both the Conservative and Labour parties.

The UK is not alone in this. Reform of communications regulation has been debated in other countries, many of whom have quite recently pushed through substantial change such as the privatisation of former state monopolies or the opening-up of television broadcasting to international ownership (Stretton/OECD 2000).

Regulation must also take into account the rapid change in the context of UK general competition policy and the changing balance between sector-specific regulation and competitive market discipline. European competition law places a prohibition on anti-competitive behaviour, backed by heavy sanctions. Within converging digital media and communications the pressure of competition probably increases as it takes place both across delivery platforms or networks and between the types and range of services delivered. Open and effective competition will therefore be a key tool in the design of a new regulatory regime – but competition alone is not sufficient to deliver all public policy objectives.

Communications and media are not like other industries. They demand special attention both because of their essential role in the infrastructure of a modern economy and because of their profound cultural and democratic significance. There are, therefore, social objectives within regulation to promote public goods, such as quality and diversity of content and plurality of voice, and social inclusion, which are only partially or imperfectly addressed by competition alone. As digital technology transforms the ways in which we communicate, an entirely new debate has been sparked about how exactly these new digital media should be made to work for the widest social as well as commercial interest. Since the technology itself is neutral, it is how digital media are used by consumers, how the markets develop and how the regulatory regime encourages or constrains use which will deliver the potential benefits – and disbenefits – to society. The overall policy framework must continue to deliver the social objectives which were delivered by the old regime, such as universal service and consumer protection. What is required is an imaginative combination of old and new, with clear policy principles and well defined regulatory aims.

Main objectives

Broadly, government objectives for policies in the communication sector might be divided into economic and social/cultural objectives (adapted from Stretton 2000 p1).

Economic objectives

- Promote and sustain competition and choice as a means of minimising price and maximising quality of communication services
- Encourage investment and innovation—the public has an overriding economic interest in the development of an open, any-to-any infrastructure with maximum public access.
- Maximise the contribution of the communications sector to economic growth and performance

Social and cultural objectives

- Protection of citizens and their rights of privacy – particularly as consumer surveys indicate that lack of trust and security is currently a significant reason for people not adopting the new technology.
- Affordable universal access – specified in terms of the new technology. The major concern must be to prevent the potential social inequities of a digital divide
- Plurality of voices in content – a plural, diverse and independent media is an essential guardian of the democratic process and a range of policy instruments have encouraged this role in the past.

- Cultural diversity and national identity reflected in network content in order to build and sustain a more inclusive society.

Administrative objectives

- The regulatory objectives must also include an independent system of regulation with minimum interference from the government of the day yet still accountable, transparent, consistent, responsive and flexible. In addition, regulators generally aim to promote maximum:
  - Public involvement and consultation.
  - Speed, and flexibility of regulation.
  - Legitimacy of regulators and regulation through consistency and clear powers and goals

Convergence and regulation: what is going wrong?

Despite the immense skill and professionalism of the existing regulators, media and communication regulation is already encountering significant problems in achieving the objectives outlined above. The ongoing process of technological and market upheaval in communications exposes some of the fissures in the regulatory regime. Content and competition regulation shows signs of strain. Whilst these strains are not yet evident in open consumer dissatisfaction or explicit public protest, they are already introducing inefficiencies in the system as a whole, with impact upon economic costs and on technical and service innovation.

Content regulation and convergence

Many of the regulatory issues are most clearly highlighted in the problems apparent in the existing system of content regulation:

- Some convergent modes of delivery such as the Internet are very difficult to regulate without incurring significant costs and negative effects. While consumer self-regulation and public awareness campaigns can in the longer term replace tighter regulatory controls such as previewing, there is a danger of a shortfall in traditional regulation goals such as child protection.

- Previous balances between regulation and freedom of speech are now impossible to maintain. In the past, publicly accountable regulation was considered justifiable for those media considered most public, influential, pervasive and invasive. But new forms of delivery challenge the boundaries between public and private media. Moreover, lack of clear definition of key terms such as 'broadcast' and 'television' makes regulation of audiovisual material across convergent platforms increasingly problematic (for example, a UK company recently showed an edited version of a film on a free-to-air terrestrial TV channel whilst simultaneously promoting free access to the cut scenes over the internet).

- Simultaneously, consumers are taking more control of their own media diet through conditional access and filtering which undermines justification of freedom of speech restrictions on digital platforms. This might be seen as part of a process of effective self-regulation and developing consumer sophistication – yet a fear of harmful content inhibits people from taking up new digital media. The regulatory framework must be robust enough to promote public confidence in the new economy.

- Self-regulation lacks adequate coordination and co-regulation structures within the regulator for oversight and coordination of codes of practice developed by industry.

- Content providers across a range of formats are subject to a tangled system of regulations according to the different modes of delivery. These constitute genuine costs and restrictions on potentially high growth sectors.

- The emerging international dimension of regulation is becoming ever more significant as content is provided by vertically integrated, international companies with substantial market power and financial resources. This is in conflict with the benefits of cultural diversity, plurality of voice and local provision of content.

If the current system of content regulation is not reformed, it will very quickly become an illegitimate, seemingly repressive structure, that is wasteful of public resources and insufficient to meet consumer expectations. The processes of convergence are set to intensify rather than subside.

Competition policy and digital media

Convergence increases the scope, size and diversity of the communications industry and thereby increases competitive pressures. In the UK telecommunications market, the balance between sector-specific regulation, general competition policy and competition law has substantially shifted as the sector has grown in size and diversity, with new technologies, new services and a multitude of licensed operators. By 1995 OFTEL felt effective competition justified a lighter regulatory price control on BT (Hall, Scott and Hood 2000). In contrast, one could argue that the broadcast sector had been protected from direct commercial competition by the regulatory licensing regime until novel technology and new entrants forced a change – and this process is still being contested. The BBC presently remains independent in that it is outside the scope of regulatory controls imposed on other broadcasters and, it is often argued, this leads to significant market distortion.

The aim should be to have a stronger, more independent system of competition regulation. According to William Melody, 'If competition is going to be an effective tool of regulation that is used to help achieve both economic and social objectives, regulators will need to be proactive in ensuring that the entry doors to telecom markets remain open and competitive opportunities are maximised.' (Melody 1997b 19.)

There remains a fundamental difference between the telecommunications and broadcast sectors in the conceptual definition of 'competition'. UK general competition law was modified in the 1999 Competition Bill to incorporate European principles and thereby place a prohibition on anti-competitive behaviour, backed by heavy sanctions. The role of a sector specific regulator in competition issues then becomes defined and limited by general law and policy while the main regulatory problem becomes one of market definition. This is where sector specific knowledge and information is crucial.

Market dominance within a sector is, of itself, not a problem within the EU competition framework. In the 1999 Communications Review and July 2000 Directives, significant market power has been redefined from 25 per cent of market share to coincide with the concept of dominance as used in competition law (Stretton 2000). However, acting in an anti-competitive manner is illegal. Since the technology itself is (largely) neutral, it is how digital media are used by consumers, how the markets develop and how the regulatory regime encourages or constrains use which will deliver the potential benefits – and disbenefits – to society. The regulator should promote and encourage innovation rather than punish it. The sector regulator therefore needs both autonomy and flexible processes to encompass change and market development. In any cycle of product innovation in a network economy, there is a period of market dominance because new services lead to new markets. Competition regulation should not penalise that innovation, but have the flexibility and strength to act firmly at the right moment in market development (whilst noting the particular problems of media pluralism described above).

Border conflict: converging regulation

As one might expect, turf wars and confusions generally affect those services that appear to fall between conventional definitions. The following are some of the emerging problems with the current regulatory system:

- Conditional access systems for television were defined as a telecommunications service by the DTI (1996), (see Hills 1997 p230) which effectively marked the regulatory boundary between carriage and content. The existing structure, which involves multiple consultations between regulators and industry, is inadequate to deal quickly and decisively with foreclosure issues. In the end, the consumers suffer.
- Video on demand also straddles the border between broadcasting and telecommunications. Service providers from previously discrete sectors in broadcasting and telecommunications are now in open competition to offer video on demand to consumers by means of different appliances and network platforms. Both OFTEL and the ITC have regulatory responsibility for different bits of the same package.
- Electronic Programme Guides which, effectively, index the entire library of digital services and information are another example of a border issue. The problems are similar to those of numbering administration and number portability in the voice telephony market which caused OFTEL so much discomfort in the early

1990s. Numbering moved, within a liberalised telecoms market, from being a technical backwater to being a significant barrier to market entry. The joint ITC/Oftel consultation on EPGs has yet to report at the time of writing.

- There are some areas in which the dispersal of economic competence between regulators has mitigated against the fast and effective execution of regulatory decisions. For example, decisions about the bundling of cable channels or the unbundling of the telecoms local loop have led to legal challenges which might have been avoided had they been dealt with by a regulator with clearer guidelines, regulatory aims and focused economic competence. Even the definition of 'economic' interest varies between regulators.

- New but increasingly important issues, such as rights to access electronic public services, and regulation of interactive services such as the Internet via digital television, are not accounted for within the existing structure simply because they did not exist when the regimes were established. Serious problems, such as privacy of online personal data, currently remain invisible. Dispute resolution and complaints handling is poorly developed in most areas of the new economy. Only 23 per cent of respondents knew how or where to complain about internet services in a recent NCC survey (NCC 1999a).

- Implementation and enforcement of EU directives is another convergence issue since the UK has a different historical progression towards market liberalisation from the rest of Europe. For a decade or more, the UK was ahead in the deregulation of government monopolies but now the rest of the EU is catching up or surpassing the UK. The NRA must therefore be robust enough to influence EU policy.

What needs to be done? Key challenges of regulatory reform

Whilst the overall policy framework must continue to deliver the social objectives which were delivered by the old regime, the reformed communications regime must also deliver new social benefits such as consumer protection online and affordable access to digital connectivity. What is required is an imaginative combination of old and new, with clear policy principles and well defined regulatory aims. The regulatory regime must deliver even on inconsistent or contradictory objectives. Achieving such a balance is the result of effective political processes combined with open and inclusive methods of consultation and governance in which all stakeholders buy into the core concepts of regulation and share a common vision of the public interest. Better regulation and new public interest objectives do not necessarily mean more regulation. The aim should always be to regulate only when the objectives are clear.

General principles and goals of reform

New digital media present an historic opportunity to promote a more open, inclusive society. A focus on three main areas will help government to take active steps to develop a regulatory system that positively promotes, rather than merely reacts to problems in this transitional era:

- *Modernising regulation – removing regulation that is no longer effective or necessary and providing a more transparent, efficient regulatory structure.* Broadly, the aim of reform should be to deal with those aspects of the current regulatory system that are generally thought, by industry and consumers, to be under-performing. The five 'core principles' of effective regulation promulgated by the UK government's Better Regulation Task Force could usefully underpin discussion of reform. Regulation should be Transparent, Accountable, Targeted, Proportionate and Consistent. The implications of these principles are explored further below. There should also be clear criteria by which asymmetries within regulatory controls might be reduced or withdrawn once effective competition is established. However, in a headlong rush to modernise, the interests of those citizens, present and future, who chose not to participate in the new economy should not be ignored.
- *The public interest – updating public interest objectives to include those that are peculiar to the new digital networked communications*

environment. The regulatory regime should work within the context of general competition law, copyright and patent law, consumer law and human rights. However, there are some issues which are quite specific to the communications sectors and demand special concern and, therefore, there will necessarily be a sector specific regulator no matter the degree of convergence and market discipline. The public interest in new communications services cannot be reduced to the same public interest objectives of either telecommunications or broadcasting. For example, access to the internet and its successors or protection of consumers engaged in e-commerce are new considerations.

- *The regulatory challenges of the next decade – preventing problems that we can predict will emerge with the transition to digital communications.* Although it is impossible to predict exactly where the market and technological changes associated with digitisation will take us, it has for some time been possible to identify some areas where existing regulation is problematic and where reforms could avert further problems. This is a complex and intellectually challenging set of problems but it is essential that the government should address them in order to promote and develop the new economy, in particular in respect of three areas:

  Overlaps and inconsistency. Where there is a multitude of regulatory bodies with different foci, objectives and organisational cultures, duplication and contradiction are inevitable. Public administration and regulatory reforms are highly path-dependent. Specific histories for telecommunications, broadcasting, radio spectrum, advertising and print media in the UK imply resistance to change, to cede responsibilities, to co-ordinate or otherwise. However, this hinders an efficient regulatory process by duplicating services and possible fueling turf wars.

  Legitimacy. In a time of rapid technological change, there is danger of following the industry's lead. Legitimacy must be gained not only through statutory powers but also the wider legitimacy of effective leadership, clear definition of role and objectives, imaginative innovation in regulatory techniques, sound analysis and dissemination of information—all linking back to the principles of transparency, accountability, consistency, proportionate and targeted response.

Scope. The new media ecology warrants a thorough re-evaluation of which services should be subject to regulation, to what degree and by which means. In the current context, asymmetrical economic regulation might be optimal. New market entrants must be encouraged or even protected from anti-competitive behaviour by incumbent or dominant market players.

Competition reform

This leads to a further competition issue. Where should control over company mergers be located? Convergence, after all, is all about merging technologies and markets. Moreover, networks become more efficient as they grow. General competition law might fail to address the social and cultural public interests which guide regulation in the communications sector, in particular with regard to interconnection and media pluralism for example. This suggests that assessment of public interest should be a fundamental obligation upon any reformed regulatory body. Once again, we should consider the balance between all the stakeholders in the sector, commerce, consumers, government and citizens, and structure the statutory and administrative processes of consultation and decision making to ensure open and inclusive consultation and decision making.

In order to perform this competition function satisfactorily, the regulator clearly needs:

- sector specific expertise and sustainable financial resources
- effective statutory powers with powerful sanctions
- clear structural rules to promote competition but proportionality in response to regulatory demands
- to allow 'sufficient short-term market power and barriers to entry on the one hand to guarantee investment in new content, but sufficient technological innovation and ease of access to the market on the other to guarantee plurality and diversity and protection from the systematic over pricing or under provision to the consumer.' (Oliver 2000 p15).

- streamlined and simplified licensing procedures across the sector, including those previously seen as broadcasting markets.

What role for self-regulation and co-regulation?

There are already a considerable number of industry-based self-regulatory bodies within the communications field, most particularly in the area of content and the internet (British Board of Film Classification – BBFC, The Internet Watch Foundation, IWF). This reflects the fact that self-regulation is probably the most effective way to deal with fast-changing, technically complex areas of activity. It is also flexible and relatively cheap, though costs are passed on to consumers. Self-regulation helps solve the problem of identifying appropriate skills and knowledge within the digital sector and how to resource the appropriate staff. The government civil service is in direct competition with one of the most highly paid commercial sectors. Clearly, the demands of a rapidly convergent industry make non-state regulation (BRTF 2000) a sensible option and in a reformed regime, there is opportunity for a much broader role for such control (see Chapter 1).

Opportunity to encourage and support self-regulation and co-regulation bodies should be taken because this also links into the broader policy objectives of promoting an open, democratic and inclusive society. However, non-state regulation can only take place effectively where all the participants accept the over-riding policy and legal framework or where they can achieve consensus on cultural values and social norms. We should, moreover, be acutely aware of the price of failure. Monroe Price (2000) refers to competition self-regulation in situations of state failure as 'mafia'. Moreover, self or co-regulation is appropriate only when it is possible to guard effectively against regulatory capture by self-interest. Within the reformed regime, the relationship between state and self- or co-regulation should be developed but the appropriate balance will vary between content and infrastructure issues.

International regulation?

New global media challenge national regulations. On the other hand, telecommunications has been addressing precisely such problems since the first Atlantic cable was laid. Co-operative intergovernmental groups are common in this sector dealing with issues such as standards, satellites and spectrum management. There is a role, however, for a sector specific regulator as expert advisor to the government in international negotiations. There is a need for a strong and coherent voice within government and one with an international profile to address the problems of international carriage, content and competition issues specific to the communications industry and also world trade issues. The current fragmentation of regulation exacerbates these problems.

Proposals for reform

IPPR recommends setting up a single regulatory authority for communications: Ofcom. Ofcom should have responsibility for regulating across the digital communications industries.

Many have argued that there are significant dangers in the promotion of a single institution with responsibility for regulating accross the communications industries. They argue that content issues will be lost if they conflict with economic issues and that a regulator would be too powerful. These concerns are valid, but not insurmountable. They should not be given as reasons to duck the responsibility of reforming a regulatory system that is in difficulty. Such dangers should be seen as challenges for institutional design, rather than reasons to stay with an outmoded and clearly failing regulatory structure.

Whilst a piece of this length cannot and should not provide institutional blueprints, the aims and principles for regulatory reform can be outlined. The following are some goals of reform, together with suggestions of means to achieve them:

- *Improve speed of decision by regulator*
  - appoint a single DG, with an advisory board
  - promote self-regulation in context of policy framework
  - single content review and rating on demand body

- *Improve strength of regulator*
  - clear definition and statement of aims and objectives in regulation
  - effective and adequate statutory powers including information gathering powers

  - statutory requirement for publication of industry guidelines and reasons for decisions

- *Improve expertise*
  - sustainable and adequate financing such as proportionate levy from the industry
  - flexible staff recruitment such as secondment from industry
  - promote industry self regulation wherever appropriate

- *Improve legitimacy*
  - establish adequately resourced Communications Consumer Council
  - statutory requirements for consultation such as public hearings, industry and consumer forums
  - promote dissemination of research and regulatory analysis
  - clear separation of functions: day-to-day regulatory decisions
  - strengthen independence from government

- *Improve transparency*
  - establish single complaints handling body with high public profile and public hearings
  - single authority for negative content regulation (taste and decency, classification).

- *Deliver new public interest objectives such as universal access*
  - a new regulatory department for Universal Service and Public Access, linked to government electronic service delivery agencies

- *Reduce political involvement in regulatory decisions*
  - clear separation of policymaking and regulatory implementation
  - DG appointed for fixed term free from interference

Some ongoing controversies

- *Organisation and leadership.* It is clearly important that the regulator should be able to react quickly and have swift sanctions which are effective and consistent. There are two particular concerns with the design of the regime; should it be a nominated individual person as the regulator or should it be a commission? Secondly, to whom and when and how should the regulator be accountable?

- *One leader or a commission?* The advantages of appointing an individual person to be a regulator are independence, speed of decision making, flexibility and efficiency and as a single, accountable point of contact. The potential problems include erratic and idiosyncratic decision making, unconstrained powers and a lack of public accountability. By contrast, a collective commission is seen to be more accountable and able to make better, more reasoned decisions and to maintain independence by avoiding regulatory capture. A sensible solution would ensure the advantages and combine the potential advantages of the technologies. In practise, the commissioners will end up making individual decisions with the rubber stamp of other members whilst the individual regulator is obliged to work in a collegiate manner and consult widely.

  Within the UK, where policy has traditionally been led by Government and Parliament and enforcement is passed to another body such as a regulatory agency, the problem of regulatory capture has not been a significant problem. In terms of practical decision making, a more radical change would be to increase and formalise wider public and user involvement in decision making, increasing the scope and depth of consultation rather than perpetuating the old pattern of consultation solely within the great and good. Such novelty is possible precisely because of the open, democratic nature of the digital communications.
- *Accountable to whom?* Again, concern for formal patterns of reporting and supervision or sponsorship are a diversion. What is of greater effect is which other stakeholders have a formal statutory role or which hold informal influence within the regulatory architecture. Which voices are excluded? It is here that, again, greater voice for consumers and citizens must be incorporated into the design of the reformed regulatory regime so as to deliver the social and cultural objectives of diversity, plurality, quality content and the

promotion of an inclusive society. There are clear advantages in terms of administrative simplicity and bureaucratic speed in making a regulator report to a department, to a minister or to Parliament. However, this must not be permitted to alter the balance between economic and social/cultural regulatory objectives and the focus upon public interest.

- *Financial structure.* Regulator finance has implications across a range of concerns: regulators independence, its ability to attract the necessary talent and expertise, and its capacity to deal with challenging issues quickly and decisively. Clearly, finance is crucial. Self regulation does have costs, they are simply passed on to the consumer. Current regulators are financed directly by industry (BBFC) or by a levy on the broadcasting license (ITC) or by government grant and receiver licence fee (BSC, BBC board).

- *Checks and balances.* We want a powerful regulator that is able to exert swift and clear sanctions in order to promote a competitive, plural, inclusive media ecology in the UK. So power itself is not the problem. This is a problem of encouraging accountability, transparency and proportionality, and of separating the policymaking function from the implementation function. Public involvement should be maximised through a consumer consultation body, as the NCC (National Consumer Council) and the CA (Consumers Association) have proposed (NCC 1999b). Policy should be transparent and made by the sponsoring ministry, whilst the regulator focuses on more neutral issues. Another check on power could be a right of appeal of regulatory decisions, and preliminary decisions should be given quickly and subject to approval. As Colin Scott put it: 'Debates over accountability have to grapple with the uncomfortable dilemma of how to give sufficient autonomy to these actors for them to be able to achieve their tasks, while at the same time ensuring an adequate degree of control' (Scott 2000 p39).

  Subjective decisions, particularly on content issues, should be supplemented with public involvement such as the use of a consumer council and also public opinion polls.

- *Balancing the public interest.* Some claim that content issues need to be housed with competition in order for them to be adequately dealt with. Others claim that competition and content need to be separate in order for content issues not to be the poor neighbours of the more tangible economic issues. Clearly these issues need to be unpacked with an eye to how regulation is transparent and accountable, and the forms of licensing and sanctions that it will involve. The concern that content would be a poor relation in regulatory decision-making is a valid one. The challenge is to devise a system that ensures content issues receive the same attention as economic issues.

  The debate about whether the three clusters of functions: content, carriage and economic regulation should be housed under one roof is in fact of secondary importance. It is crucial that balance of importance in terms of content, economic and carriage issues must be maintained. Where actions (such as license renewals, spectrum allocation, mergers decisions) involve consideration of issues across that range of issues, there should be no horse trading, and clear guidelines regarding transparency and public involvement when content and other (such as interconnection, competitiveness) concerns conflict.

  This could be done through clear procedures for publication of reports of public interest tests and reports for licensing, in a way that made public the foundation for every decision. Even if the regulator has one single name and structure, it is possible for the content department to publicly call for sanctions from the licensing department. Ultimately, economic sanctions have to be available to content regulators.

The process of reform

Achieving the goals outlined above will not be easy, partly because media regulation is such a politically sensitive area. The IPPR preference is for fair competition and general competition law to have an increasing role and for oversight regulation to be reduced and removed where appropriate in both carriage and content. However, public interest and consumer protection must remain paramount in the reformed architecture and function of the new regime.

The key problem is how to manage the process of regulatory reform. The detailed architecture of Ofcom should be the result of consulation and involvement with existing regulators.

Two matters are of paramount importance:

- The need for clear leadership from government
- The need for in-depth consultation and buy-in from regulators.

Some Key Features of Ofcom

General

- Improved international operations and legitimacy
- Department of universal access working on electronic delivery of public services
- Co-regulation capacities, agreeing codes of practice with industry
- Independence to develop transparent policy in consultation with public.
- 2 Pillar structure within regulator, and separate board meetings for public hearings when content and other issues clash.
- A degree of functional separation of content and economic issues: An internal pillar structure separating 'subjective and expert' or content and economic issues.
- Communications council for public involvement in key decisions
- A larger panel/board with more public involvement should deal with more subjective issues. A board of five technical regulators should deal with economic issues. These latter should be full time appointees.
- Issues of conflict between content and economic boards should be dealt with publicly.
- Checks and balances: clear separation of policy and implem entation functions. Independence to *implement* economic/carriage policy without industry/ government interference.[1]
- A single address for public complaints.
- Statutory status. In order to ensure independence and guard against regulatory capture, the regulator should be a statutory body.

Competition

- Sector specific behavioural competition regulation with goal of promoting content pluralism and public access
- Continuing public interest tests for mergers with oversight of audience share, and specific responsibilities for media pluralism.
- A regulator with sufficient sanctions and resource to deal with new sites of market foreclosure, CAS, EPG, and ex-post/ behavioural approach to dominance and pluralism.

Content

There is a broad consensus that content regulation will remain important in the medium to long term, that it will rightly remain an issue dealt with at the level of the nation-state and that the current fragmented structure is unsustainable. However, we are moving towards an environment in which content issues will be increasingly dealt with throughindustry self-regulation. Since uniform ratings and structures cannot be applied across all delivery systems, and the sheer volume of content is beyond any control, it is likely that content regulation will be dealt with on a flexible scale (see Chapter 4).

Negative content regulation (Taste and decency, safety)
The ITC and the BSC should be immediately merged, and take on the functions currently performed by the Internet Watch Foundation. They should take on the additional task of rating interactive content (eg games) that is specifically designed for children, on a complaints driven/ regulation-on-demand model working with the BBFC.

- Because of the differing degrees of invasiveness of services, and consumer expectation, core mass services will continue to require oversight, but other channels will not. Content regulation should be applied according to a single sliding scale (see Chapter 4)
- Those services not obliged to rate should be able to access a rating-on-demand service for self-regulation accross platforms
- Improved transparency and public profile of regulator due to increased importance of notice and take down and reliance on public complaints
- Need for single ratings service for content exploited across multiple platforms
- Need for improved coordination and oversight of co-regulation arrangements and codes of practice

Positive content regulation (Quality, positive content requirements)

- A regulator should apply a single sliding scale to all services according to consumer expectation and market share, and financial/spectrum priveliges afforded to broadcasters.
- The single regulator should be financed by a combination of licensing, industry contribution and taxation.
- The single content regulator should regulate the BBC. Particular attention should be paid to the issue of regulating political impartiality in the BBC and other public broadcasters with duties to respect due impartiality.
- Public involvement and opinion research should be built in to the structure of content classification and review, and progressively replace the 'great and the good model.

1 This point is developed further by Melody (1997a,b)

Leadership from government

The aims of regulatory reform should be kept constantly in mind. These are to:

- Prevent existing regulation constituting a barrier to convergence
- Improve speed, cost and efficiency
- Remove unnecessary regulation

- Improve transparency and accountability of regulators

Need for in-depth consultation and buy in from regulators

Since any change in public administration is very path dependent, one key issue is institutional culture. It is well recognised that when commercial companies merge they usually become substantially less profitable for a period before the benefits become evident. This time lag is not permissible within a government body and the regulator must be effective from the outset. Given the very different organisational cultures and the different focus of work within the current regulatory bodies, amalgamation should be undertaken with care and consultation, and with a clear notion of common objectives. During the process of reform, The DGs of regulators must meet regularly in a board with statutory status: a shadow strategic communications authority. This should meet and make reccommendations on the model of the Shadow Strategic Rail Authority.

Clearly there will be tradeoffs and balances to be struck, but adherence to a combination of the general principles adhered to by the Better Regulation Taskforce and the specific principles outlined in this pamphlet will ensure that the reformed regulator performs its role.

Summary

IPPR recommends setting up a single authority for communications regulation – OFCOM – to replace the ITC, the BSC, Oftel and ICSTIS, and incorporate functions currently carried out by the BBFC and the IWF.

The advantages of a single regulator include:

- Better access to scarce knowledge and expertise of industry
- Strength, resources, swiftness of action
- Linking of content and economic issues
- Efficiency through economies of scale, particularly regarding international liaison; research, public involvement, and public complaints, and through merging of research functions and resources

The detailed blueprint for a new institution should not be decided in advance, but arise from a detailed process of consultation and cooperation between existing communications regulators, government, the public and interested parties, taking into account the dangers and goals of reform outlined above.

Summary of recommendations

Broadcasting is changing. Fast. Regulators need to keep apace of these changes, and there is a responsibility for governments to ensure that citizens are served well despite the changes afoot. IPPR media has argued that significant reforms are necessary across the field of media and communications regulation. The articles contained in this volume, and the other products of the IPPR Media programme 1999-2000 have outlined numerous policy recommendations. Amongst those recommendations are the following:

- *OFCOM* Merging the fragmented communications regulators into a single structure, which will take responsibility for regulating competition, carriage and content across the communications industries. This will replace regulators that were previously defined in terms of technology, such as telephony, television or film.
- *Content* Applying a single guiding framework for content regulation across various media, utilising a framework based on a combination of the market share of a service, and the consumer expectation of regulation.
- *Competition* Ownership rules should continue to be applied to analogue broadcasting services, but may eventually be relaxed if
  - a new competition regime is designed with measures to promote pluralism at its centre and
  - if media democratisation reforms
- *Self-regulation* There will be an increased role for self regulation and co-regulation in the future. The regulator should be charged with developing codes of practice in partnership with industry, and with promoting public awareness of self regulation by consumers using filters.
- *Access* Extending universal access to information services should be a key role of Ofcom. This should include monitoring access to public services delivered electronically.
- *The BBC* There must be separation of the commercial businesses from the public service operations with completely transparent trading between them taking account of quantified value of the

brand and of licence fee investment. BBC services should be regulated by OFCOM, with a reduced role for the governors.

More detailed recommendations are listed in the individual chapters of this book. Together, they offer a compelling case for reform and some outlines for the key directions that government policy should take if public authorities are to continue to protect the public interest in communications into the digital century.

Bibliography

Aitman D (1994) 'Cross Ownership of Media Interests in the European Community' Speech to AIC Conference on Multimedia and Broadcasting Reform 29-30 September

Arnback J (2000) Regulation for Next Generation Technologies and Markets *Journal of Telecommunications Policy* 24 (Online)

Balkin JM, Noveck BS and Roosevelt K (2000) 'Filtering the Internet – A Best Practices Model' in Waltermann J and Machill M (eds) *Protecting our Children on the Internet* pp199-261 Guetersloh: Bertelsmann Foundation

Baumol W and Sidak J (1994) *Towards Competition in Local Telephony* Cambridge: MIT Press

Beard D, Cave M and Holmes J (2000) 'Ownership Issues in Broadcasting from a Competition Policy Standpoint' Unpublished seminar paper London: IPPR

Benkler Y (1998) Communications infrastructure regulation and the distribution of control over content *Journal of Telecommunications Policy* 22 pp183-196

Blackman CR (1998) Convergence between telecommunications and other media *Journal of Telecommunications Policy* 22 pp163-170

Blinderman E, Price M and Verhulst S (2000) *Codes of Conduct and Other Self-Regulatory Documents: Emerging Patterns of Norm Formulation and Enforcement on the Internet* Bertelsmann Foundation

Blumler J (1986) *Television in the United States: funding sources and programming consequences* London: HMSO/Home Office

Burkert H (2000) The Issue of Hotlines in Waltermann J and Machill M (eds) *Protecting our Children on the Internet* pp263-318 Guetersloh: Bertelsmann Foundation

Cabinet Office Better Regulation Taskforce Homepage (www.cabinet-office.gov.uk)

Camesasca PD (2000) Mayday or Heyday? Dynamic Competition Meets Media Ownership Rules after Premiere *European Competition Law* Review 2 pp76-87

Cavallin J (1998) 'European Policies and Regulations on Media Concentration' *International Journal of Communications Law and Policy* 3 (online http://www.digital-lawnet/IJCLP/)

Cleevely D 'Scenarios for Convergence and the Internet: Communications in 2015' in *E-Britannia: The Communications Revolution* 2000

Clements B (1998) 'The impact of convergence on regulatory policy in Europe' *Journal of Telecommunications Policy* 22 pp197-205

Collins R, Garnham N and Locksley G (1987) *The Economics of Television* London: Sage

Collins R and Murroni C (1996) *New Media, New Policies* London: Polity Press/IPPR

Collins R and Purnell J (1995) *Managing the Information Society* London: IPPR

Collins R and Purnell J (1996) *Reservoirs of Dogma* London: IPPR

Corry D (1995) *Regulating in the Public Interest* London: IPPR

Corry D (ed) (1994) *Regulating Our Utilities* London: IPPR

Council of Europe Committee of Experts on Media Concentrations and Pluralism (MM-CM) (ed) (1997) Strasbourg: Council of Europe

Curran J (2000) 'Press reformism 1918-1998 – A study of failure' in Tumber H (ed) *Media Power Professionals and Policy* pp35-55 London: Routledge

Curran J and Seaton J (1991) 'A Palette of Policies' in Curran J and Seaton J (eds) *Power Without Responsibility The Press and Broadcasting in Britain* Chapter 21

DCMS (1999a) *Creative Industries: UK television exports inquiry* London: Department of Culture Media and Sport

DCMS (1999b) *Regulating Communications – The Way Ahead* Results of Consultation Paper (at 114) London: Department of Culture Media and Sport

DCMS (1998) *Select Committee on Culture Media and Sport – Fourth Report* House of Commons 520

De Sola, Pool A (1974) 'The Rise of Communications Policy Research' *Journal of Communication* pp31-42

Department of National Heritage (1995) *Media Ownership The Government's Proposals* CM 2872 London: HMSO

Department of National Heritage (1994) *The Future of the BBC* London: HMSO (Cm 2621)

Department of Trade and Industry (1999) *Review of the Internet Watch Foundation* HMSO

Dertouzos J and Trautman W (1990) 'Economic Effects of Media Concentration: Estimates from a Model of the Newspaper Firm' *Journal of Industrial Economics* 39 pp1-14

Doyle G (1997) ' From 'Pluralism' to 'Ownership': Europe's Emergent Policy on Media Concentrations Navigates the Doldrums' *Journal of Information Law and Technology* (JILT)

Elton M (1992) 'The US debate on integrated broadband networks' *Media Culture and Society* 14 pp369-396

European Commission (2000) 2257th *Session of the Telecommunications Council* Brussels, EC

European Commission (1999a) *Results of the Public Consultation on the Convergence Green Paper: Communication* [Com(1999)108]

European Commission (1999b) *Internet Action Plan* European Commission

Feintuck M (1998) *Media, The Public Interest and the Law* Manchester: Manchester University Press

Feintuck M (1997) 'Regulating the Media Revolution: In Search of the Public Interest' *Journal of Information Law and Technology* (JILT)

Garnham N (1999) 'Barriers to Convergence' in: Dutton WH (ed) *Society on the Line: Information Politics in the Digital Age* Oxford: Oxford University Press

Gibbons T (2000) 'Pluralism Guidance and the New Media' in Marsden C (ed) *Regulating the Global Information Society* London: Routledge

Gibbons T (1998) *Regulating the Media* London: Sweet and Maxwell

Gibbons T (1992) 'Freedom of the Press: Ownership and Editorial Values' in *Public Law* Summer 1992 pp279-299

Goldberg D P T and Verhulst S (1998) *Regulating the changing media: a comparative study* Oxford: Clarendon Press

Graham A (2000) *The Future of Communications* Oxford: Oxford University Press

Graham A and Davies G (1997) *Broadcasting Society and Policy in the Multimedia Age* Luton: University of Luton Press

Graham D *et al* (1999) *Building a Global Audience British Television in Overseas Markets* London: Department of Culture Media and Sport

Grewlich K W (1999) 'Cyberspace: Sector-Specific Regulation and Competition Rules in European Telecommunications' *Common Market Law Review* 936-969

Hall C, Scott C and Hood C (1999) *Culture Chaos and Interdependence Inside the Regulatory Process* London: Routledge

Hoffman-Riem W (1996) *Regulating Media: The licensing and supervision of broadcasting in six countries* New York: The Guildford Press

Home Office (1986) *Report of the Committee on Financing the BBC – The Peacock Report* London: HMSO

Hotelling W (1929) 'Stability in Competition' *The Economic Journal* pp41-52

Hughes J (2000) 'The Global Marketplace: Making Sense of the Future' in *E-Britannia: The Communications Revolution* Luton: University of Luton Press

Humphreys P (1994) *Media and Media Policy in Germany: The Press and Broadcasting Since 1945* Oxford: Berg

Humphreys P (1996) *Media and Media Policy in Western Europe* Manchester: Manchester University Press

Humphreys P and Lang M (1998) 'Digital Television between the Economy and Pluralism' in Steemers J (ed) *Changing Channels: The Prospects for Television in a Digital World* pp73-96 John Libby Media

Hutchison D (1998) *Media Policy: an introduction* London: Blackwell

Independent Television Commission (1999) *Television: the public's view* London: ITC

Independent Television Commission (1997) *ITC Code of Conduct on Electronic Programme Guides* London: ITC

Iosefides P (1999) 'Diversity versus Concentration in the Deregulated Mass Media Domain' *Journalism & Mass Communication Quarterly* 76 52-162

Iosefides P (1997) 'Methods of Measuring Media Concentration' in *Media, Culture and Society* 19 pp643-663

Keane J (1992) *Democracy and the Media* London: Polity Press

Kruger D (1999) *Access Denied* London: Demos

Levy D (1999) *Europe's Digital Revolution: Broadcasting regulation the EU and the nation state* London: Routledge

London Economics/Office of Fair Trading (1994) *Barriers to Entry and Exit in UK Competition Policy* London: Office of Fair Trading

Magaziner I (2000) 'The Role Governments Should Play in Internet Policy' in Waltermann J and Machill M (eds) *Protecting our Children on the Internet* pp61-78 Guetersloh: Bertelsmann Foundation

Markoff J and Montecinos V (1993) 'The Ubiquitous Rise of Economists' *Journal of Public Policy* 13 pp37-68

Marsden C (1999) *Pluralism in the Multi-Channel Market: Suggestions for Regulatory Scrutiny* European Council

McGregor O (1977) *Report of the Royal Commission on the Press* London: HMSO

McQuail D (2000) 'Media Policy: Premature Obsequies?' in Tumber H (ed) *Media Power, Professionals and Policies* pp19-35 London: Routledge

McQuail D (1992) *Media Performance, Mass Communication and The Public Interest* London: Sage

Media Metrix (2000) Europe Over A Quarter Of UK Households Have Access To The Internet (http://www.mediametrix.com)

Melody WH (1997b) *Policy Objectives and Models of Regulation* Lyngby: Technical University of Denmark

Mohamed SH (2000) *New Regulatory Framework for Communications and Multimedia Sector: Subsidiary Legislation* Malaysia: Malaysian Communications and Multimedia Commission

Murroni C, Collins R and Coote A (1996) *Converging Communications – Policies for the 21st Century* London: IPPR

Murroni C and Irvine N (1998) *Access Matters* London: IPPR

Napoli P (1999a) 'Deconstructing the Diversity Principle' *Journal of Communication* 49(4) pp7-34

Napoli P (1999b) 'The Marketplace of Ideas Metaphor in Communications Regulation' *Journal of Communication* 49(4) p151

National Consumer Council (1999a) *Ofcom – A coherent approach to regulating communications* NCC

National Consumer Council (1999b) *Tuning in to Consumers* NCC

Nihol P (1998) Competition or regulation for multimedia? *Journal of Telecommunications Policy* 22 pp207-218

Noam E (1993) 'From the Network of Networks to the System of Systems: An End of History in Telecommunications Regulation?' *Regulation* 16 (Online)

Oftel (2000) *Review of universal telecommunication services: A consultative document issued by the Director General of Telecommunications*

Ogus A (1995) 'Rethinking Self-Regulation' *Oxford Jounal of Legal Studies* 15 pp97-108

Oliver M (2000) *Scenarios for Convergence and the Internet – Implications for Content and Content Providers* London: Department of Trade and Industry

Pilkington H (1962) *Report of the Committee on Broadcasting 1960* London: HMSO

Policy Action Team (PAT) 15 (2000) *Closing The Digital Divide: Information and Communication Technologies in Deprived Areas* London: Department of Trade and Industry

Price M (1998) *The V-Chip Debate Content Filtering From Television to the Internet* Lawrence Erlbaum

Price M (1994) 'The market for loyalties: Agenda setting for a global communications commission' *Intermedia* 22 pp14-21

Price M and Verhulst S (2000) 'The Concept of Self-Regulation and the Internet' in: Waltermann J and Machill M (eds) *Protecting our Children on the Internet* pp133-198 Guetersloh: Bertelsmann Foundation

Prosser T (1997) *Law and the Regulators* Oxford: Clarendon Press

Raboy M (1998) 'Public Broadcasting and the Global Framework of Media Democratization' *Gazette* 60 pp167-180

Raboy M *et al* (1998) 'Global Media Policy – A Symposium on Issues and Strategies' *The Public* 5 pp63-105

Robertson G and Nicol A (1992) *Media Law* 3rd edn London: Penguin

Rogers P (1997) 'The Case for the Single Regulator' *Spectrum: The Quarterly Magazine of the Independent Television Commission* 24 pp6-7

Samarajiva R (2000) 'Establishing the legitimacy of new regulatory agencies' *Journal of Telecommunications Policy* 24 pp183-187

Saundry R and Nolan P (1998) 'Regulatory Change and performance in TV production' *Media Culture and Society* 20 pp409-426

Scannell P (1996) *Radio Television and Modern Life* Oxford: Blackwell

Scannell P (1989) 'Public Broadcasting and the Global Framework of Media Democratization' *Media Culture and Society* 11 pp135-166

Scott C (2000) 'Accountability in the Regulatory State' *Journal of Law and Society* 27 pp38-60

Stretton A O (2000) *Changing Approaches to Regulation of the Communications Sector: Some Lessons From OECD Countries* Report Prepared for IPPR seminar 'Reforming the Regulators' July 18 2000:

Tambini D (2000a) *Rethinking Media Content Regulation: A Rolling Programme of Reform* London: IPPR (unpublished)

Tambini D (2000b) *Digital Danger* London: IPPR

Tambini D (2000c) *Promoting Pluralism* London: IPPR (unpublished)

Tambini D (2000d) *Convergence and UK Creative Industries: Flexible Strategy for Future Export Performance* London: IPPR (unpublished)

UNESCO (1999) *Public Action: Regulation or Economic Incitement* Paris

Ypsilanti D and Xavier P (1998) Towards next generation regulation *Journal of Telecommunications Policy* 22 pp643-659